GLIMPSES
INTO
THE LIFE OF INDIAN PLANTS

GLIMPSES
INTO
THE LIFE OF INDIAN PLANTS

AN ELEMENTARY INDIAN BOTANY

BY

I. PFLEIDERER

FOURTH EDITION

GYAN

Published by

GYAN

Gyan Publishing House
5, Ansari Road
Daryaganj, New Delhi-110002
Phone: 011-47034999, 9811692060
E-mail: books@gyanbooks.com

Distribution Network
gyanbooks.com
India, USA, Canada, UK, Australia, France

ISBN : 978-81-212-3433-7 (PB)
First Published, 1921

2nd Impression 2020

Printed at: Gyan Press, Delhi.

The book at your disposal is the edition of an old book originally published long back. The Text quality of this book may not be at par as compared to our other books but many readers approached us to reprint this important book so that it could satiate their hunger. We request all the readers of this book to help us to improve its quality by providing the necessary text of those pages which are not properly visible. Your valuable support in this case will be highly appreciated and acknowledged and our future readers will be benefitted immensely.

GLIMPSES

INTO

THE LIFE OF INDIAN PLANTS

AN ELEMENTARY INDIAN BOTANY

BY

I. PFLEIDERER

FOURTH EDITION

PUBLISHED BY THE
KANARESE MISSION PRESS AND BOOK DEPOT, MANGALORE
1921

PREFACE TO THE FIRST EDITION

SOME years ago I was asked by a number of friends in Mangalore to help them in the study of Botany. The promise given at that time led me to compile this little book, which, I hope, will rouse their interest in the life of plants they were so anxious to know.

At the same time I was prompted to come forward in writing a simple book on Indian Botany by the sad disappointment I felt when I witnessed the poor and lifeless way in which object-lessons on plants were taught in our elementary schools. I understood that, if any improvement in this subject was to be made, the teachers themselves had first to be interested in nature-study. And as a treatment of Botany in the old way could not secure the object in view, I adopted a course which I was glad to find in DR. O. SCHMEIL'S Manual of Botany, intended to open to any student of ordinary intelligence an understanding of plant-life and to enlarge and quicken in him a sense of that infinite harmony which runs through every part of the Creator's marvellous plan of nature, which would make the educational value of this subject equal to that of any other subject taught in schools.

To this end I combined the structural description of plants with a plain description of their vital processes. The technical terms, which so often form the crux of beginners, are reduced to the smallest number possible, and many statements are illustrated by suitable cuts. The illustrations may also be supplemented by the coloured plates of Indian Plants, published by the Kanarese Mission Book and Tract Depository,

Mangalore, at a very moderate price, to which reference is made in the text, wherever possible, and of which the eight coloured pictures found in the book, are reduced reproductions.

I need hardly emphasize that in using this manual as a school-book the plants studied are to be put into the student's hand, and that the types described should, if the climate permits, be planted in a school garden for the continued observation of the various stages of their life. Many of them may, at least in the West Coast of India, be procured in any field, wood, or garden.

The book does not aim at completeness, which is not required in a school-book. Yet, the natural orders selected are systematically grouped and the characters of the various classes and divisions are briefly stated, so as to give the student at once an insight into the classification of the vegetable kingdom. . . . . .

Mangalore, 1908. *I. P.*

PREFACE TO THE SECOND EDITION

The present little volume has in its first edition been more favourably received by the public than I ever expected. I ascribe this success more to the growing interest in the subject than to the merits of the book itself. Above all, I beg to offer my sincere thanks to the educational departments in the South and North of India for the kind recognition of my work.

The printing of a new edition has been taken advantage of to make some corrections and to introduce new matter, so as to increase the usefulness of the book. If I have erred on the side of excess, I trust my readers will find it easier to eliminate than to insert.

I am told that the types selected to represent the several families are not common to all parts of India. My personal knowledge of the Indian flora being more or less limited to the West Coast, I do not know whether it would be possible to find out type-species suiting all parts of this great Peninsula. I must, therefore, leave it to my readers to substitute other types from the number of genera and species mentioned in each family.

For the identification of plants I recommend any one of the books on the Flora of India named in the bibliography, page 280. Nairne's *Flowering Plants of Western India* is particularly useful for residents of the West Coast.

The nomenclature and classification have not been materially changed and are those of BENTHAM and HOOKER.

In using this book for school purposes I beg to offer the following suggestions to teachers:—

(1) The book does not contain a scheme of lessons, but should be used as a storehouse from which the teacher may draw.

(2) The following curriculum may, perhaps, suggest itself to teachers of High Schools in Southern India:

First year.—Descriptions of individual plants, such as Lotus, Cotton tree, Mango tree, Bean, Cucumber, Sunflower, Bindweed, Chillies, Tumbe, Banyan tree, Cocoanut palm, Kesu (Taro), Gloriosa, Banana, Rice, Fern.

Second year.—Structure and Life of Plants: Leaf, Root, Stem (pp. 198—251).

Third year.—Floral Biology: Flower, Fruit, and Seed (pp. 251—276).

Fourth year.—Factors determining the life and structure of plants:—

The Water: Roots and their functions.—Transpiration.—Hygrophytes.—Xerophytes.—Tropophytes—Water and Marsh plants.

The Light: Assimilation.—Thirst for light.—Climbers.—Protection from intense light.

The Soil: Whether porous or not, whether manured or not.—Geological formations.—Mangrove.—Flora of dunes.—Epiphytes and Parasites.—Saprophytes and Carnivorous plants.—Symbiotic plants.

The Air: Mechanical and physiological effects of the wind.—Strength of Stems.—Rarification of air in high mountains.

I am deeply convinced that Botany is pre-eminently the branch of Science most fitted for the young. And if by putting forth this book I have done anything to rouse the interest of those who teach and those who learn for this noble and pleasurable study, I shall have been largely rewarded for my efforts.

UDIPI,
October 31st, 1911.

I. Pfleiderer.

PREFACE TO THE THIRD EDITION

One is glad to know that there is an increasing demand for this book and that the second edition is exhausted.

It is evident that a large number of students are learning Botany by the aid of this book. I hope they will find that the many alterations made by the author in this edition will not only greatly assist them in their study, but also make this subject even more attractive.

HASSAN, MYSORE STATE,
September 1916.

A. Brockbank.

TABLE OF CONTENTS

FIRST PART

DESCRIPTION OF TYPES AND FAMILIES

SECOND PART
THE STRUCTURE AND LIFE OF PLANTS

Appendix.

FIRST PART

DESCRIPTION OF TYPES AND FAMILIES

DIVISION I.

FLOWERING PLANTS (Phanerogamæ).

All plants in this division have stamens and pistils, and form seeds.

CLASS I.—DICOTYLEDONS

Plants with two seed-leaves. Leaves net-veined. Parts of flower usually in sets of five or four. Stems, if woody, consisting of a woody substance growing in circles round a central pith, and surrounded by bark.

SUB-CLASS I.—POLYPETALÆ

Plants generally with two floral envelopes: calyx and corolla. Petals separated from one another.

1. The Water-Lily Family

(Nymphæaceæ).

Aquatic herbs. Flowers radial. Petals and stamens numerous, inserted on the receptacle. Ovary superior, syncarpous, *i. e.* formed of (many) cohering carpels.

The Lotus Water-Lily *(Nymphæa lotus).*

(Plate No. 623.)*

(*Kan.* Naidile, Tāvare. *Mal.* Veḷḷānpal, Nīrāmpal. *Tam.* Veḷḷāmbal, Indirāvāčam. *Tel.* Allikāḍa. *San.* Sītōtpala.)

Many quiet tanks and peaceful lakes are adorned by the beautiful Lotus flower. The broad leaves spreading over the

* Plate No. 623 of Mangalore series of Coloured Pictures of Indian Plants and Trees.

surface of the water like floating shields, and the lovely flowers, like large white floating lilies, increase the mysterious charm that all still waters possess in the sunshine. No wonder that many myths have arisen about this plant, and that it is almost worshipped by many nations!

Most plants soon die, if their roots and stems are kept under water. Not so the Lotus; for it lives there, and its structure is wonderfully adapted to a life in such surroundings.

1. The **Stem** is thick and tuber-like, with many scars which show where the leaves formerly grew. It creeps along the ground, sending forth shoots at its upper end and decaying at the other. Such a stem is called a root-stock or rhizome. It sends down stout, long roots into the soft mud so that it may not be carried away by any movements of the water in which it grows.

2. From the stem rise the long stalked **Leaves,** which so long as they are below the surface are rolled up with their edges inward. As soon as the leaves reach the surface, the stalk ceases to grow, and the leaves unroll their broad blades for the sun and air to play on. To these, like all other green plants, the Lotus must have access in order to feed and breathe. When the water rises to a higher level, the stalks stand vertically; when the level sinks lower, they move more and more sideways, like the ribs of an umbrella which is being opened.

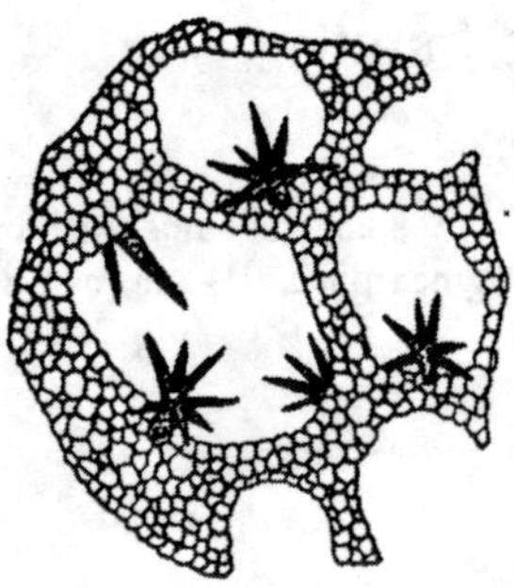

Fig. 1.—Transverse section of the leaf-stalk of the Water-lily with large air-chambers and bristles within.

If one of the leaves is torn off the stem, it will float. This is due to air-chambers in the stalk, which can be easily seen if the long stalk is cut across (fig. 1). The chambers contain bristles, which serve as a means of protection against the voracious water-snails, whch would otherwise feed on the leaf-stalk and so destroy the leaves.

The upper side of the leaves is covered with a wax-like substance, so that any water which may fall on them runs off, as it would from a duck's back. We may also notice that the middle of the leaf, where the stalk joins it, is a little higher than its edges which are slightly waved. This helps the water to run off. If the water were to remain on the leaves, it would hinder the growth of the plant by stopping up the little holes (stomata), through which the plant absorbs carbon dioxide, a gas which, with the aid of water, the plant is able in the sun-light to turn into starch, and thus gain the carbon which forms an essential element in all plants.

The openings in land plants are more numerous on the lower side of the leaves, but in the lotus they are all on the top, because the lower side of the leaf rests on the water, and they would be of no use on that side. Put the blade of a leaf under water and blow down the stalk, and you will soon see little silvery bubbles form on the top-side of the leaf: this is the air escaping through the stomata. The air can thus communicate with the root-stock and the roots. This is important for the life of the plant. Every living part of a plant must be accessible to the oxygen in the air. Now, the muddy soil in which the Lotus grows, has scarcely any oxygen, as it is used up in decomposing the dead vegetable substances deposited there. It is through the air-channels in the leaf-stalks that the life-sustaining air can penetrate down to the very root of the plant.

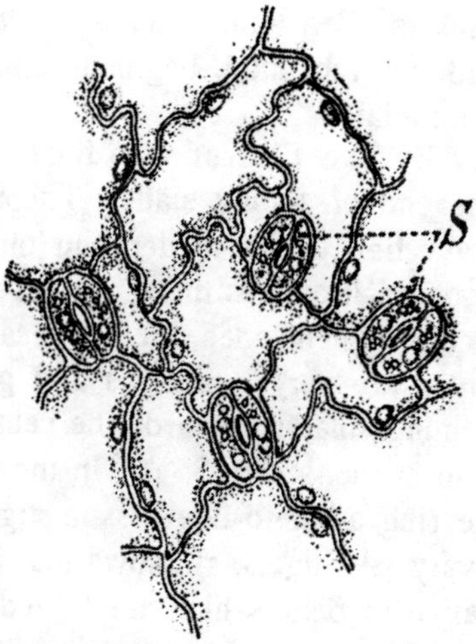

Fig. 2.—Part of the epidermis of a leaf (200 times enlarged). *S.* Stomata.

Another point about the leaves is worthy of notice: the lower side is of a darker colour than the upper, the reason for this being that dark coloured things absorb heat better than light coloured ones. (A black coat is much hotter in the sun than a white one.) Now the heat rays which accompany the light rays from the sun would pass through the leaves into the water, if they were bright green throughout; but the violet colour absorbs the heat rays and thus assists the growth of the plant. Increase of heat causes quicker growth.

When the tank in which the Lotus grows dries up, the leaves with their long stalks sink down in the mud and perish. The plant, however, does not die with the leaves. For, when the tank is filled with water again, the root-stock (stem), which was hidden in the mud, begins to sprout again, and the plant is thus perennial.

3. The **Flower** also floats on the surface of the water at the end of a long stalk. The four green sepals form a good protection to the tender bud on its journey to the top, and as soon as it arrives there, the sepals open out looking like small boats. Their inner side is the same colour as the petals. These latter are very numerous and grow in a spiral, gradually becoming smaller towards the centre and at the same time turning into stamens (fig. 3, 2). In the centre is the ovary or seed-box bearing a shield-like sessile stigma. A transverse section of the ovary (see fig. 3, 5) shows that it is composed of many leaves called carpels, which are folded in towards the floral axis and thus form several cavities in which the seeds ripen.

The flowers open after sunrise, and are visited by insects which are attracted by the colour. They, however, only find pollen, as the plant secretes no honey. Towards evening the flower closes again to protect the delicate stigma and the pollen from damp and cold.

4. The **Fruit,** a spongy berry (fig. 3, 4), ripens below the surface of the water. When the seeds are ripe and leave the berry, a small bubble of air under the slimy mantle (called aril) that covers the seed, brings them to the surface, and the seeds are carried wherever the wind and waves take them until the bubble bursts. The seed being heavier than water sinks to the bottom and begins to grow to form a new plant which may be at some distance from the parent one. In this simple way the Lotus plant is enabled to spread.

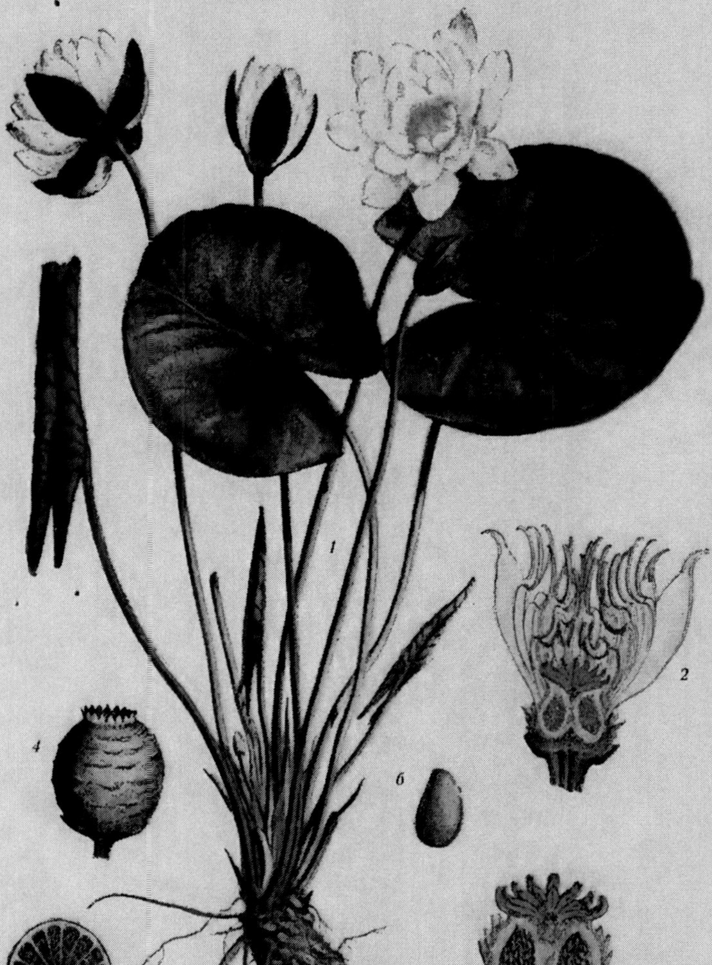

Fig. 3.—WATER-LILY *(Nymphæa lotus)*.

2. Vertical section of flower. 3. Vertical section of ovary. 4. Capsule. 5. Transverse section of it. 6. Seed.

Other Water-Lilies.

There is a red variety of the Lotus, *viz., Nymphæa lotus var. rubra* (*Kan.* Kenneidile, Kendāvare), which is the **Egyptian Lotus.** The smaller sort of Nymphæa, *N. Stellata*, has red, white, or blue flowers. The **Sacred Lotus** of India is *Nelumbium speciosum* (*San.* Padmā), the funnel-shaped leaves and the large rosy flowers of which are raised over the surface of the water. The root-stock and the seeds are eaten.

2. The Custard Apple Family

(Anonaceæ).

Leaves alternate, simple; stipules 0. Flowers radial. Sepals three, petals three or six. Stamens numerous. Carpels numerous, free. Seeds large, endospermous.

The Custard Apple Tree *(Anona squamosa).*

(Plate No. 653.)

(*Kan.* Sītāphala. *Mal.* Āttaċakka. *Tam.* Āttāmaram. *San.* Subhā.)

1. **Home.**—The Custard Apple is cultivated in gardens in the whole of India for its sweet, delicious fruit. Its original home is the West Indies.

2. **Stem and Leaves.**—The plant has a woody stem branching from the ground. Such a type of plant is called a shrub. Its branches are long and slender and bear alternate leaves in two rows to right and left (bifarious). The leaves are short-petioled, simple, oblong, entire, and glabrous. When crushed they have an aromatic smell.

3. The greenish **Flowers** grow single from the axils of the leaves and are small and, therefore, little conspicuous. The calyx consists of three small, triangular sepals, and the corolla of as many thick, oblong petals, alternating with the sepals and valvate in the bud. In the centre of the flower the receptacle is thickly set with numerous short-stalked stamens in its lower and outer part, and with numerous carpels, each with its short

style, in its upper and inner part. If a single stamen is pulled out and examined with a hand lens, a thick cap will be noticed on the top of the anther. This is an extension of the tissue (called connective) that connects the two halves of the anther.

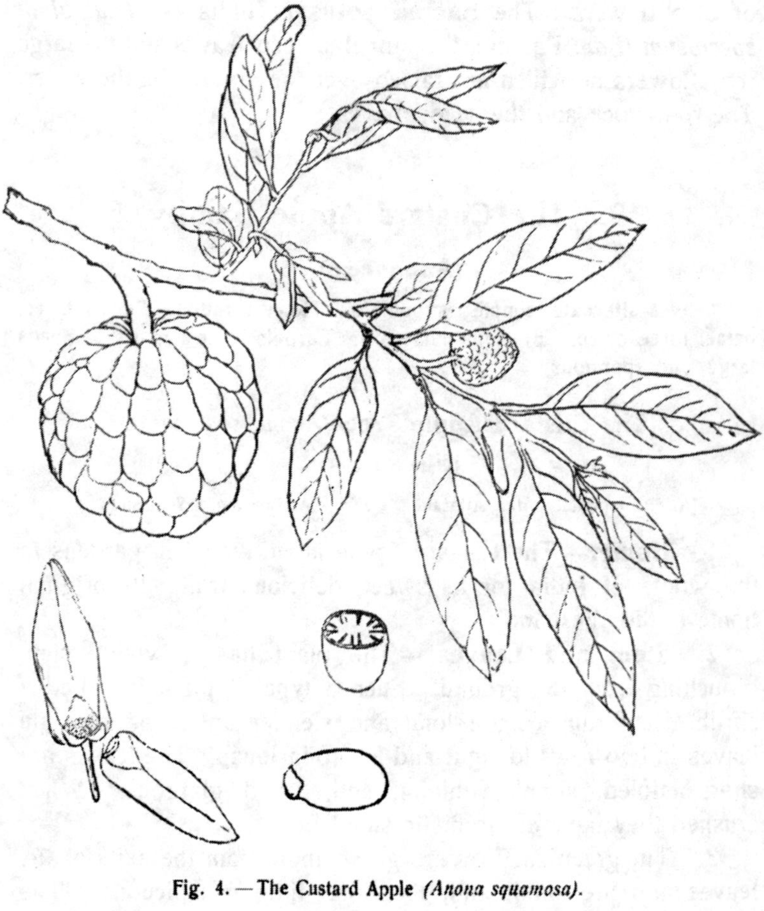

Fig. 4. — The Custard Apple *(Anona squamosa)*.

It may also be noticed that the anthers are on the outer side of the stamen. (Such stamens are termed extrorse. The stamens of the Lotus are introrse.)

4. The carpels are free from one another (apocarpous) in the flower, but they are confluent in the **Fruit,** which represents an aggregate of many fleshy carpels, densely packed together and each containing one seed. The outer side of each carpel becomes hard in the fruit and forms one of the parts of the hard protective crust. The seed has a hard testa which projects into the interior, the greater part of which is filled with endosperm, a nourishing tissue for the embryo required at the time of its germination.

A wild growing type of this family is *Uvaria narum* (*Kan.* Unāmiṇi; *Mal.* Narampāṇal; *Tu.* Pāṇḍel), a straggling climber with branches dividing from the zigzag-like young shoots at angles of nearly 180°, and thus forming hooks by means of which the climber fastens itself to its support. The numerous carpels are stalked and free and become scarlet when ripe.

3. The Poppy Family

(Papaveraceæ).

Herbs with milky juice. Flowers radial. Sepals two, petals four, stamens many, inserted on the receptacle. Ovary superior, syncarpous. Seeds endospermous.

The Opium Poppy *(Papaver somniferum).*

(*Kan.* Kasakase. *Mal.* Kaçakaça. *Tam.* Gasagasā. *Tel.* Gasagasa. *San.* Khaskhasa.)

The Opium Poppy is widely cultivated in India, because it yields *"opium"*, which is an important article of commerce. Opium is a valuable drug which can soothe the greatest pain and cause sleep. It is obtained by scratching the unripe capsules (fruit) with a small knife. The milky juice which comes out is allowed to dry and is then scraped off the capsule. In China this substance is smoked in a pipe for intoxicating purposes. The opium smoker soon sinks into a half-conscious state in which pleasant dreams come to him. On waking up again he is, however, very dull and suffers from headache; and in order to get rid of the discomfort he again indulges in a smoke. After much indulgence, the dose must be increased for the desired effect, and this slowly but surely undermines the health of the smoker until he sinks into an untimely grave.

1. The **Stem** and **Leaves** have a bluish coat of wax on them, which, as we have seen in the case of the Lotus plant, serves as a means of protection against the choking up of the little breathing holes (stomata) by rain water. The leaves near the root are larger than those higher up. If it were not so, the upper leaves would prevent the sunshine from falling on the lower ones and make them useless for making starch. The upper leaves also lie closer to the stem: this serves the same purpose. Besides, having the large leaves low down, the stem need not be so strong; and the leaves also afford shelter to the root and keep the soil there damper than if they grew at a height from the ground.

If the stem is wounded, a milky juice comes out. This dries and hardens and so protects the tender cells inside from further injury. The juice also gives the whole plant a peculiar smell and taste, which animals do not like and so prevents them from feeding on the leaves.

2. The **Flowers,** when in bud, have a calyx of two green sepals. These drop as soon as the flower opens, and the petals, which were crumpled up inside, open out and become quite smooth, and the flower with its four large, shining, white petals becomes very conspicuous. This plant, too, secretes no honey, yet it is visited by insects that eat the pollen which is produced so abundantly on the numerous stamens. Whilst eating they scatter the pollen dust about, covering themselves. The scattered pollen is caught by the large shell-shaped petals and kept ready for the next insect visitor. The upright position of the flower also helps to prevent the pollen from being spilt and wasted on the ground.

The petals are too weak to carry the weight of the heavy insects that come to visit the flower; another resting place is, therefore, provided, namely the broad-rayed stigma that spreads over the ovary like a shield. The insects which have visited other poppies and got covered with pollen alight here and naturally drop some of the pollen-grains thus fertilising the ovules or future seeds in the ovary (fig. 5, 4).

3. If the **Fruit** is cut across (fig. 5, 3), the structure is easily seen. There are numerous walls inside, like the radii of a circle, which, however, do not quite join in the centre, and it is on these walls that the seeds grow, until they are ripe when they drop off into the space between the walls. In order to let them escape from the capsule (as the fruit is called), little windows open all round the top just below the stigma (fig. 5, 5;

Fig. 5. — 1. Opium Poppy *(Papaver somniferum)*. 2. Diagram of flower. 3. Transverse and 4. Vertical section of ovary. 5. Ripe capsule. 6. and 8. Seed. 7. Stamen.

poricidal) and if the plant is shaken by the wind, the tiny seeds are thrown about in all directions. Bend down one of the ripe capsules and let it go, and you will see what happens. We now see also why the stalk is tall and springy, as the higher the seeds are from the ground the further will they be scattered.

Other Poppies.

The **Mexican Poppy** *(Argemone Mexicana)* is a well-known weed with prickly white-veined leaves and a yellow juice in all its parts. Oil for lamps is extracted from the seeds. The seeds are also used as a purgative.

Fig. 6.—Corn Poppy *(Papaver Rhœas)*.

The **Corn Poppy** *(Papaver Rhœas)* is sometimes grown in gardens for its beautiful scarlet flowers. To the farmer the Corn Poppy is a pest, for it robs the corn of nourishment, light and space.

4. The Crucifer or Mustard Family

(Cruciferæ).

Herbs without milky juice. Flowers radial, with four sepals and four petals, arranged crosswise. Stamens six, the two outer shorter, and the four inner longer. Ovary superior, formed of two carpels. Fruit a pod, divided into two valves by a central frame to which the seeds adhere.

The Indian Mustard *(Brassica juncea)*.

(Compare Rape, Plate No. 625.)

(*Kan.* Sāsive. *Mal.* Kaḍuka. *Tam.* Kaḍugu. *Tel.* Āvālu. *San.* Sarshapah, Raktajāji, Dundubha.)

If you bruise some Mustard seeds between two pieces of paper, a grease spot is left on the paper. This is due to the presence of a fatty oil in the seed. Some plants, like the Rose and "Tulasi", have another kind of oil in their leaves which does not cause a grease spot, but vanishes quickly. This kind of oil is called a "volatile" oil and is generally the cause of the scent in flowers. The "fatty" oil serves as a food for the young seedling.

The Mustard seed has also a volatile oil which is very pungent and will cause tears to come to the eyes. It is this oil which protects the seed from being eaten by birds, and which makes the seed useful to us as a condiment or medicine, and it is for the oils in its seeds that the plant is so commonly grown.

1. The **Stem** grows to a height of 4 or 5 feet and spreads into many branches.

2. As in the Poppy, the **Leaves** gradually become smaller as they grow higher up the stem. In this case, though, the upper leaves have quite a different shape from those which grow at the bottom: the upper leaves have no stalks and are long narrow and toothed, whereas the latter have long stalks and are lyrate, that is, are lobed and have the end lobe larger than the others. (The leaves of Rape are stem-clasping, while those of Mustard are not: see Rape, Plate No. 625.)—The leaves, too, point upwards; and if you watch the rain falling on them, or

pour water on them, you will see that the water is taken to the stem and runs down to the root.

3. The **Root** system of the Mustard plant, instead of spreading to a distance in all directions, like that of, for instance, the Mango tree, forms a distinct tap-root with a few thin side-roots only. It is for this reason that in the Mustard plant the leaves carry the water to the centre of the root system.

4. The **Flowers** of the Mustard plant are bright yellow and have four sepals and four petals. The latter are stalked and grow inside, but alternate with the sepals. There are six stamens, two with short filaments, and four (inside) with longer ones (fig. 7). The centre of the flower is formed by an ovary with a short style. If cut across (see diagram, fig. 8), it can be seen that it is made of two carpels, the edges of which each bear a row of ovules.

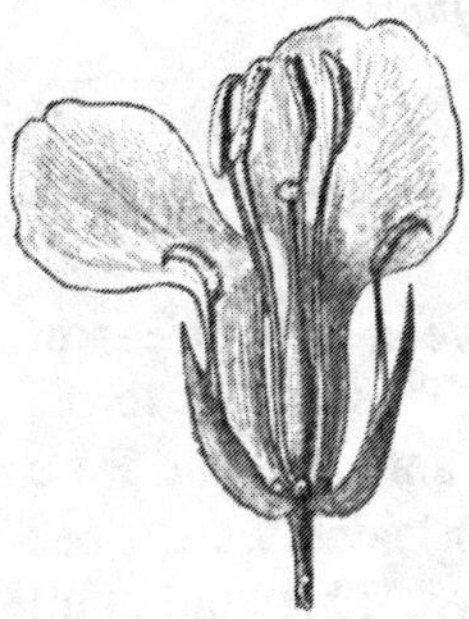

Fig. 7.—Flower of Mustard (*Brassica juncea*). One sepal and two petals are removed. (3 times natural size.)

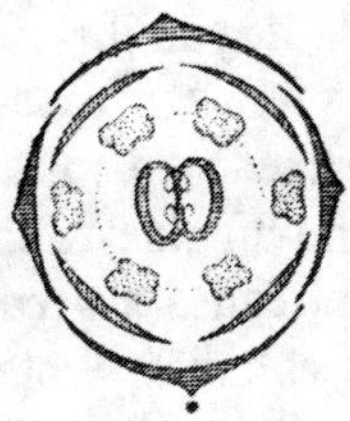

Fig. 8.—Diagram of a Crucifer flower.

The bright colour of the flowers, which open at the same time, and in large numbers, attracts the passing insects to look for the honey which is there at the base of the stamens. The insects pay for the kindness of the flower in providing them such a nice drink by spilling pollen on the stigma of the style and so fertilising the seeds.

5. The **Fruit** is a long erect pod formed of two dry carpels which split upwards and outwards from the base, showing the seeds growing on a central frame (fig. 9). Such a pod is called a silique.

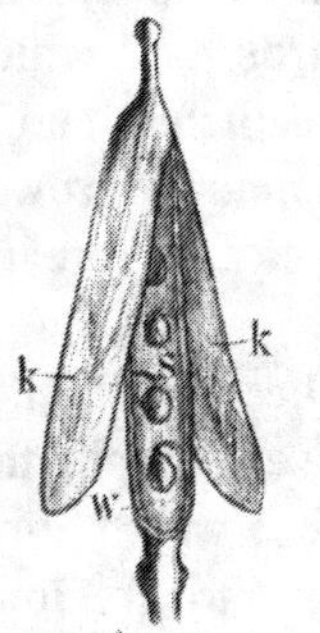

Fig. 9.—Silique of the Mustard plant. *k.* Carpels. *W.* Central frame. *S.* Seeds. (Natural size.)

Other Crucifers.

The family to which the Mustard plant belongs, contains a

very large number of plants in cool climates, but there are not many genera of this family in India. It takes its name from the crosswise arrangement of the different parts of the flower. Amongst representatives of the family we may mention the **Radish** (*Raphanus sativus,* Plate No. 625, 7, 8), the swollen root of which is eaten; the **Rape** or Coleseed (*Brassica napus,* Plate No. 625, 1–6), from the seeds of which oil is made, and **Cabbage** *(Brassica oleracea)* from which, by cultivation, all the varieties of cabbage, cauliflower and knolkohl have been produced.

5. The Mallow and Cotton Family.

(Malvaceæ).

Herbs, shrubs, or trees. Leaves alternate, stipulate, often palmate. Hairs usually branched (stellate). Flowers radial, showy, with each five sepals and petals, the latter twisted in bud and overlapping. Stamens many, united into a staminal column or into five bundles, and adnate at the base to the petals. Anthers one-celled and kidney-shaped. Ovary superior, syncarpous. Fruit a schizocarp, or a capsule.

The Indian Cotton *(Gossypium herbaceum).*

(Plate No. 629.)

(*Kan.* Araḷe, Hatti. *Mal.* Karuparutti. *Tam.* Parutti. *Tel.* Patti. *San.* Kārpāsah.)

Several varieties of Cotton are cultivated in India, which are probably referable to three main species:—*a)* The Tree Cotton *(Gossypium arboreum), b)* the American Cotton *(G. Barbadense),* and *c)* the Herbaceous Cotton *(G. herbaceum).*

1. **Stem** and **Leaves.**—The Herbaceous Cotton is a perennial and bushy shrub in the warmer areas, and annual where the cold weather being severe kills the plants. The stems are erect, the branches spreading, and the leaves alternate, stipulate and half segmented into three or five or seven broad lobes.

2. **Flowers.**—In their axils the flower-buds grow and are protected not only by the cup-shaped calyx, but also by three large heart-shaped bracteoles forming an involucre. (Plate 629, 4.) The five large petals are bright yellow, usually rendered

more conspicuous by being coloured dark purple at their base. The numerous stamens, so combined with the petals that they form a tube covering the ovary and the style (fig. 11, 3), produce large quantities of yellow pollen-dust for the insects to feed on. The anthers (which are best examined in flower-buds) are one-celled and kidney-shaped. The long style grows through the tube, formed by the stamens, and bears five stigmas above everything else, and they are therefore in the best position to catch the pollen brought by insects from any other flower they have visited. Without pollen the fruit will not set, and the plant will not bear any seed.

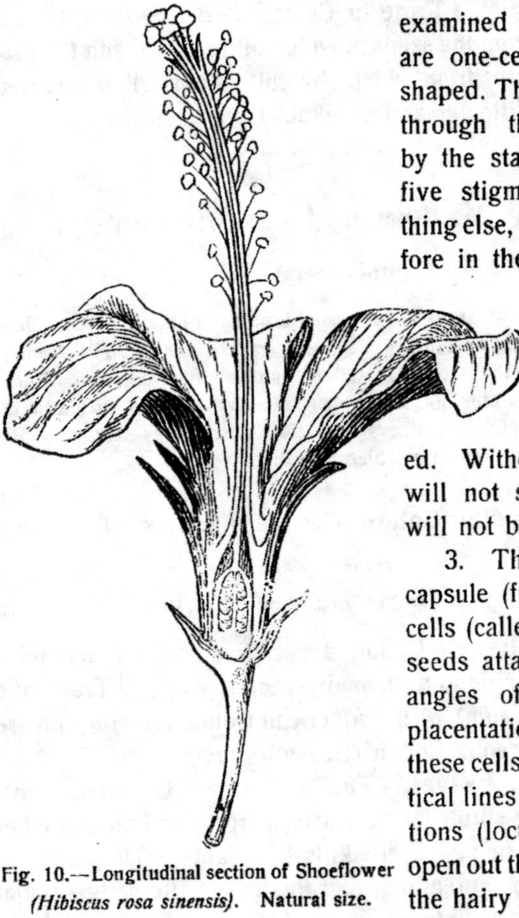

Fig. 10.—Longitudinal section of Shoeflower *(Hibiscus rosa sinensis)*. Natural size.

3. The **Fruit** is a dry capsule (fig. 11, 4) of three cells (called *loculi*) with the seeds attached to the inner angles of each cell (axile placentation). When ripe, these cells burst by three vertical lines between the partitions (loculicidal), so as to open out the cells and expose the hairy seeds to the action of the wind which may carry them to a distance. These hairs are "cotton", and it is on account of them that the plant is so largely cultivated in almost all tropical countries. The cotton is gathered by hand, dried in the sun, passed through a gin to remove the seeds, and then pressed

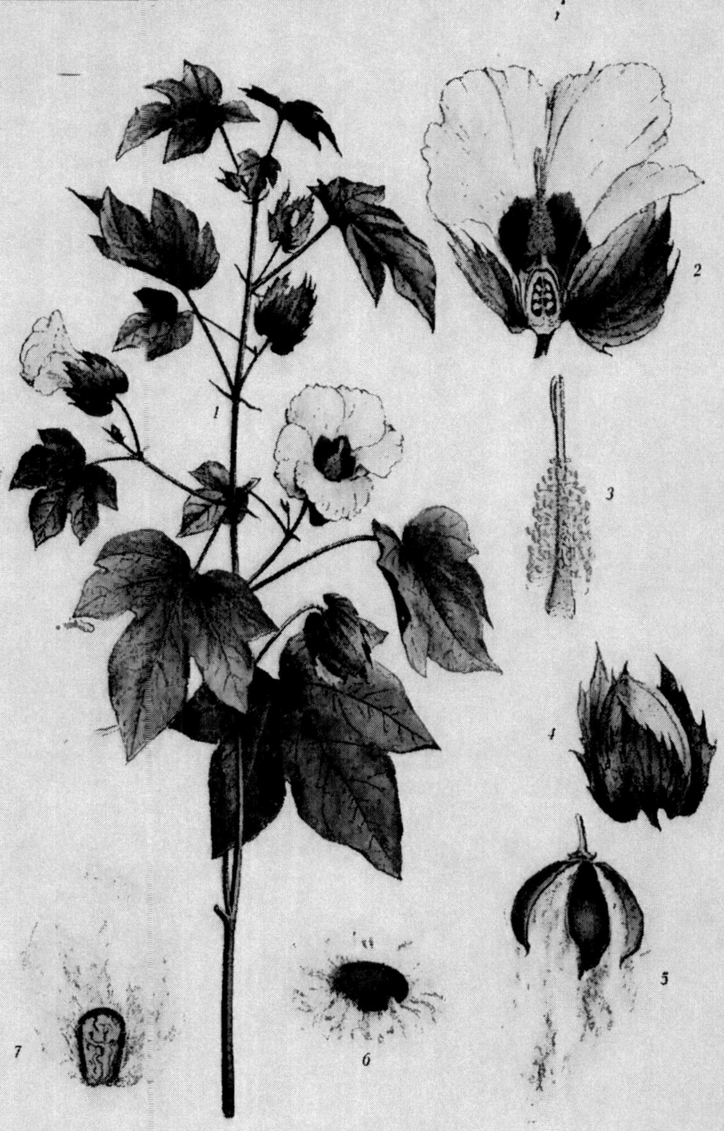

Fig. 11.—Herbaceous Cotton (*Gossypium herbaceum*).

2. Vertical section of flower. 3. Monadelphous stamens. 4. Unripe capsule. 5. Ripe capsule. 6. Seed. 7. Section of seed.

by machinery into very hard bales. The bales are sent to the mills, where the cotton is spun into thread and woven into cloth.

4. The cotton plant is also **useful** in various other respects. The stems of the plant yield a good fibre. The seeds (fig. 11, 6 and 7) are in many parts of India thrown away as a useless article. But they can be given to cattle, especially to milch cows, to increase the flow of their milk, for cotton seeds contain an oil which is nourishing.

Fig. 12.—*Bombax malabaricum*, stripped of its leaves in the dry season.

Other Mallows.

Many Indian plants belong to this family, as the **"Ladies' Fingers"** *(Hibiscus esculentus)*, the **Shoeflower** (*Hib. rosa sinensis; Kan.* Dāsāḷa; *Mal.* Čemparutti; *Tam.* Šembarttai; *San.* Japa; fig. 10),—the **Portia Tree** (*Thespesia populnea; Kan.* Huvarasi mara; *Mal.* Pūparutti; *Tam.* Pūvarašu; *San.* Kundaḥ), the **Red Silk Cotton Tree** (*Bombax malabaricum; Kan.* Kempu būraga; *Mal.* Muḷḷilavu;

Tam. Ilavu; *San.* Pūraṇī), and the **White Silk Cotton Tree** (*Eriodendron anfractuosum; Kan.* Biḷi būraga; *Mal.* Pūḷa). The garden shrub *Hibiscus schizopetalus* has hanging flowers with ciliated petals; in *H. mutabilis,* **Changeable Rose,** the flowers change from white to red in the course of the day; **Roselle** (*H. sabdariffa)* has a red fleshy calyx from which jelly is made. All of these have capsular fruits, like the Cotton plant. Common wild Mallows are **Sida** *carpinifolia* (*Kan.* Vishakaḍḍi), *S. rhombifolia, S. cordifolia,* **Urena** *sinuata,* **Abutilon** *indicum,* all of which have indehiscent carpels which, however, separate from the axis of the fruit when ripe. Such fruits are called schizocarps.

Allied Families.

The **Chocolate Tree** (*Theobroma cacao.* Plate No. 627) belongs to the *Sterculiaceæ,* a family nearly allied to the Cotton family. The Chocolate tree was introduced into India from tropical America. Its gourd-like fruits contain, in their sour pulp, very bitter seeds, the so called Cacao-beans (fig. 13, 3. 4). These are cleaned, roasted, and powdered, and then become Cacao or Cocoa, which, mixed with sugar and flavoured with Vanilla, makes Chocolate.

Another allied family is the *Tiliaceæ,* of which we mention the **Jute** *(Corchorus capsularis),* a small shrub largely cultivated in Bengal for its fibre which is manufactured into coarse fabrics, such as gunny-bags, the common coarse bags in which the various grains are sent to market. It grows wild also here and there in Southern India, and is easily recognisable from its globular wrinkled capsules that resemble Rudrakshi beads.

6. The Tea Family

(Ternstrœmiaceæ).

Trees or shrubs. Leaves alternate, simple, generally leathery, stipules 0. Flowers radial. Sepals and petals each five imbricate. Stamens multiples of five. Ovary superior, three to five-celled.

The Tea-shrub *(Camellia theifera).*

(Plate No. 624.)

(*Kan.* Cāgiḍa also "Teagiḍa". *Mal.* Cāyaccaḍi. *Tam.* Thay-ila.)

The Tea-shrub is found wild in the jungles of Assam, but is

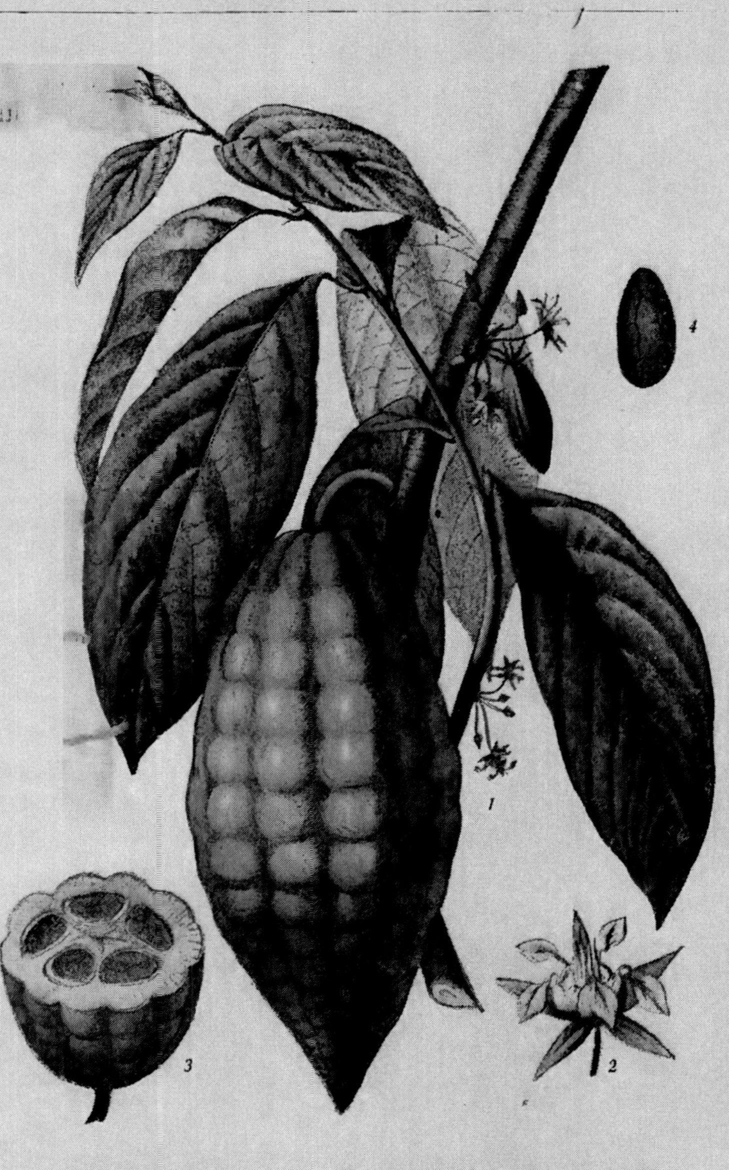

Fig. 13. — CHOCOLATE TREE *(Theobroma cacao)*.

2. Flower. 3. Transverse section of fruit. 4. Seed.

extensively grown on the slopes of the Himalayas, the Nilgiris, other mountains, and in Ceylon. It has leathery, shining leaves of a dark-green colour with toothed edges, which reflect the hot rays of the sun and thus keep the tree cool. It flowers all through the year and bears beautiful white flowers with five or more large petals and numerous yellow stamens, adhering partly to the base of the petals and partly to the ovary (Fig. 14, Plate No. 624, 2. 3). When ripe, the woody capsule opens in such a way (Plate No. 624, 7) that each of the three carpels splits in the middle (loculicidal), thus letting the seeds escape.

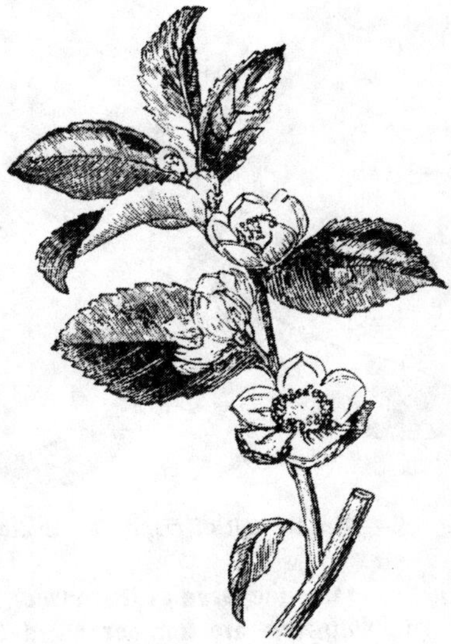

Fig. 14.—Flowering branch of the Tea-shrub *(Camellia theifera)*.

The parts from which the beverage "tea" is made, are the leaf-buds and the two or three young leaves next the buds. The volatile oil which is the cause of the flavour of tea and an alkaloid, called "theine", which has a soothing effect on the nerves, are contained in the leaf-buds more than in any other part of the plant. These leaves are picked carefully from the shrubs, partly dried in the sun or by machinery, rolled, and finally roasted to complete the drying. Of these dried leaves an infusion is made with boiling water, and this has the same flavour and odour as the tea leaves themselves; if allowed to stand for a long time, the infusion (tea) becomes spoiled, because a substance, called tannin, which is very astringent and is injurious to digestion, is dissolved. Tea must, therefore, be drunk soon after the infusion is made.

Allied Families.

Allied families are the *Dipterocarpeæ* to which some very common trees of our forests belong; *viz.*, the **Sal Tree** (*Shorea robusta; Kan.* Āsina mara; *Mal.* Sāla; *Tel.* Sālamu); the **White Dammar** (*Vateria indica; Kan.* Dhūpada mara; *Mal.* Payannumaram; *Tam.* Veḷḷaikunnrikam); **Hopea** *parviflora; Kan.* Bōvu; *Mal.* Irubōgam; *Tam.* Koṅgu); and *Hopea Wightiana; Kan.* Karmara, Hiribōgi; *Mal.* Illapoṅgu; *Tam.* Karam). In these trees (*Vateria* excepted) all or some of the lobes of the calyx enlarge after flowering and act as wings for the fruit. *Hopea Wightiana* is attacked by an insect producing in the axils of the leaves a fruit-like growth which contains the larva of that insect.

Fig. 15.—Winged fruit of *Hopea Wightiana.*

The *Guttiferæ* are another allied family, members of which are the **Pinnay Oil Tree** (*Calophyllum inophyllum; Kan.* Surahonne; *Mal.* Ponnakam; *Tam.* Punnai), and **Garcinia** *indica* (*Kan.* Puṇarahuḷi, Murginahuḷi; *Mal.* Punampuḷi).

Of the *Violaceæ* (Violet family) we mention the pretty little *Ionidium Suffruticosum* (*Kan.* Purusharatna; *Mal.* Ōrilattāmara), the **Sweet Violet** (*Viola odorata*) and the scentless **White Violet** of the Ghats (*V. serpens*).

7. The Orange Family

(Rutaceæ).

Trees and shrubs, with alternate leaves dotted with transparent glands. Flowers radial. Sepals and petals four or five. Stamens joined at their base into various groups. Ovary superior, syncarpous.

The Citron (*Citrus medica*).

(Plate No. 630.)

(*Kan.* Mādavāḷa. *Mal.* Māduḷaṅgam. *Tam., Tel.* Mādiphalamu. *San.* Mātulaṅga).

The fruit of this tree is a berry with about ten divisions

(cells) under the cover of a thick leathery skin which, in its outer part, contains numerous glands of aromatic oil (Plate No. 630, 3. 4). The sour pulp in which the few seeds lie, consists of swollen hairs filled with juice. The aromatic skin as well as the sour pulp are much prized.

Fig. 16.— Floral diagram of Citrus.

The fruits grow in the leaf-axils of evergreen shrubs or small trees. Their leaves are alternate and elliptical and have serrate edges. If you crush them, they smell strongly; for they are filled with that volatile oil which we have already noticed in the skin of the fruit. Hold a leaf up to the light, and you see its whole blade dotted with oil-glands.

Fig. 17.— Flowers and leaves of the Orange tree (half size).

The white and fragrant flowers consist of a cup-shaped, five-toothed calyx, a corolla with five fleshy petals which soon drop after unfolding, and numerous stamens whose broad filaments are joined into various bundles round the pistil (fig. 16).

From *Citrus medica* various varieties have been produced by cultivation: so the **Lemon** *(Citrus medica var. limonum)*, the **Sweet Lime** *(C. med. var. limetta)*, and the **Sour Lime** *(C. med. var. acida)*.

Other plants belonging to this family are the **Orange** *(Citrus aurantium; Kan.* Kittaḷe; *Tam., Mal., San.* Nāraṅgam), the **Pummelo** *(C. decumana; Kan.* Čakōtra, Sakkarekañji; *Mal.* Madhuranārakam), the **Bael Tree** (*Aegle marmelos; Kan.* Belapatre; *Mal.* Kūvaḷam; *Tam.* Vilvam), the tree **Zanthoxylum** *Rhetsa* (*Kan.* Jimmi, Kāvaṭe; *Mal.* Cuyiṭṭi), and **Murraya** *Kœnigii* (*Kan.* Karibēvu).

Allied Families.

There are various plants belonging to allied families which cannot be fully described here, but deserve a passing notice.

To the *Lineæ* belongs the common **Flax Plant** (*Linum usitatissimum*; *Kan.* Atasi; *Tam.* Aḷiviral; *San.* Atasi). This is cultivated

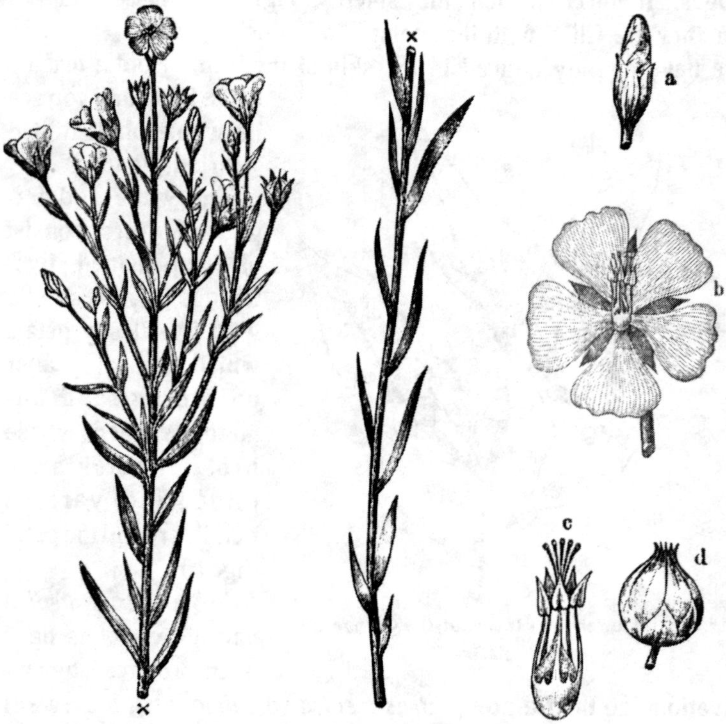

Fig. 18.—Common Flax *(Linum usitatissimum)*.
a. Flower-bud. *b.* Flower. *c.* Stamens and pistil. *d.* Ripe capsule.

throughout India for the oil contained in its seeds (Linseed oil), and also for the fibres of its stem. The slender stems bear alternate, small leaves and pretty blue flowers, collected at the ends of the branches (fig. 18). The flowers are composed of each five sepals, petals, stamens (united at their base), and styles. The fruit is a round capsule, containing two oily seeds in each of its five cells.

The fibres of the inner bark or liber of the stem are very tough and can, therefore, be used for textile fabrics. To get the fibres, the plants are first stripped of their seeds and then steeped in water until partially rotten, when the gummy part of the stem will be dissolved and the fibres loosened. Next, to separate the woody portion of the stem, they are spread to dry and then "broken", by which process the wood inside becomes brittle and breaks into pieces. The fibres are then drawn through a comb, called the hackle, where they are straightened and freed from dust and other matters. The fibre, which is thus gained, has a fine, silky appearance, and is spun into yarn, and finally woven into linen cloth in the loom.

To the *Geraniaceæ* belong the following:—

The **Bilimbi Tree** (*Averrhoa bilimbi; Kan.* Bilimbi; *Mal.* Vilumbi; *Tam.* Pilimbi; *Hind.* Tamaruṅg). It grows in the yards of many houses and bears plantain-like fruits on its trunk. Its leaves are sensitive like those of many leguminous plants: they fold at night.

The Garden **Balsam** (*Impatiens balsamina; Kan.* Gauri-hū) is a very common plant during the monsoon, and its habits are characteristic of hygrophilous plants living in very moist places. The stalk and leaves are succulent, tender and covered with a bluish coat of wax (see page 8, 1). Under the tuft of leaves at the top the spurred pink flowers grow as under a protecting roof. If you pluck these pretty flowers for a bouquet, they very soon fade. As the plant grows at a time while, and in places where, water can be obtained plentifully, it is not furnished with those means which tend to check the evaporation of the sap in the plant so much required by plants living on dry soil (such as a thick epidermis, small leaf-blades, hairy surface, p. 52). So it cannot remain fresh without water and soon fades when plucked. If we touch its ripe seed-vessels, they burst with great force and cast the seeds some distance as by an elastic spring (fig. 19). The same happens, if the wind shakes the plants. Common wild Balsams are *Impatiens Kleinii* (*Kan.* Nirukaḍḍi) and *I. chinensis.* Their succulent

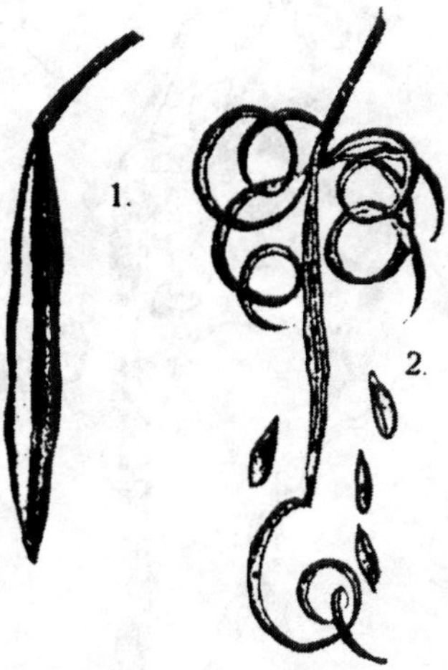

Fig. 19.—Capsule of a Balsam. 1. When closed. 2. Exploding.

stems are so transparent that the vessels conducting the sap can be seen in them as dark lines when held towards the light.

A very common tree, whose bark and leaves are in repute as medicines, is the **Neem** or **Margosa Tree** (*Melia indica; Kan.* Bēvina mara; *Mal.* Nimbam; *Tam.* Vēmbu; *Tel.* Nimbamu; *San.* Ravipriya), also belonging to an allied order, the *Meliaceæ.*

To the *Rhamneæ* belongs **Zizyphus** *jujuba* (*Kan.* Bogari), a small tree with three-nerved oval leaves in which the stipules are modified into thorns.

Fig. 20.—Branch of Grape Vine *(Vitis vinifera).*

Of the Vine Family *(Ampelideæ)* we mention

The **Grape Vine** (*Vitis vinifera; Kan.* Drākshe). This is a weak shrub which, with the help of tendrils growing opposite the leaves and homologous with the main axis terminating in them, seeks support on other plants and climbs up towards the light. It is cultivated in some parts of India and produces the sweet, juicy grapes which are spoken of as the best fruit of the whole vegetable kingdom. They are eaten fresh as dessert-grapes, or dried into raisins. The chief use, however, of this plant is the wine made from the grape. For the preparation of wine the juice is first pressed out. The sweet juice, thus obtained, soon becomes cloudy, for innumerable germs (bacteria) begin to work in it. These live in the soil of the vineyard and are blown by the wind on to the skin of the grapes, and thus come into the juice of the grapes also. Here they grow and reproduce themselves rapidly, and cause an important process, in the liquid, called fermentation. Two new substances are formed in it, namely alcohol and carbonic acid gas which from time to time bursts in bubbles and escapes. By this process the sweet juice is gradually changed into alcoholic wine. This drink has a stimulating effect on the nerves, if taken in small quantities. Its abuse, however, is very injurious to the drinker's health and is the source of much misery. For children, wine is always injurious, even if taken in very small quantities.—Another species of this genus is *Vitis quadrangularis* (*Kan.* Sandubaḷḷi; *Mal.* Caṅalamparanda; *Tam.* Ārugani; *Tel.* Vajravaḷḷi). It is common in hedges and, though a poor-looking and scraggy plant, is typical of the order. Its fleshy, cactus-like, jointed stems point to its habitat being in dry regions.

8. The Mango Family

(Anacardiaceæ).

Trees or shrubs, often with milky or acrid juice. Leaves in most genera alternate, stipules 0. Flowers radial. Sepals, petals, and stamens three to five. Ovary superior. Fruit a drupe.

The Mango Tree *(Mangifera indica).*

(Plate No. 639.)

(*Kan.* Māvu. *Mal.* Māvu. *Tam.* Mā. *Tel.* Māvi. *San.* Čūtah, āmrah.)

This tree grows all over India and is not only one of her stateliest trees, but also produces one of her best fruits.

1. The **Trunk** of the tree is covered by a dark-gray, cracked bark, when old. The young stem has a green outer skin, called epidermis, such as annual herbs have. But as the tree grows larger, the epidermis, not being able to stretch, bursts. It is now necessary for the plant to form a new protective cover, which is done by constantly forming air and water-tight layers of what are called *cork* cells. Some trees form a very thick layer of cork, like the Andipunar tree *(Carallia integerrima)*, or the Spanish Oak, the bark of which is the ordinary cork which is sold in shops. If the cork is thin, the stems have a smooth surface like the Guava, or the Jack tree. The Mango tree has a thick layer of cork which, as it thickens, cracks until flakes of bark drop off.

Under the cork layer is the *inner bark*, called *bast,* and inside this are concentric layers of wood. The younger or outer layers of wood are that part of the stem through which the sap ascends from the roots to the branches. Any injury to the bark or sapwood may interfere seriously with the circulation of the sap or let disease-fungi enter into the stem. A tree, however, has the power of healing its wounds by a rapid growth of cork at the edges which gradually cover the damaged area. This function is made use of in the process of grafting mangoes.

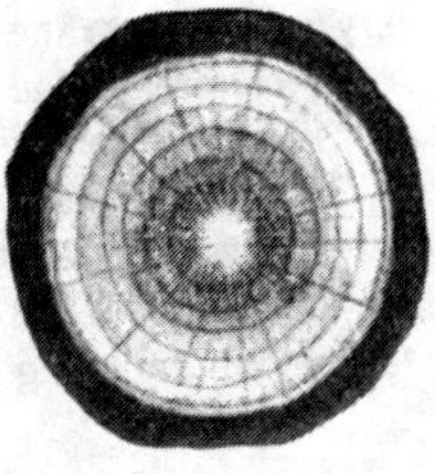

Fig. 21.—Transverse section of the stem of the Mango tree.

A small Mango plant, about as thick as one's finger, is grown from a seed in a pot, and when it is required to make a graft, a slice is taken out of one side of the stem down as deep as the pith and about an inch long. A branch of one of the good edible kinds of Mango of the same thickness is treated in the same way. This branch and the plant are so placed together that the one cut surface exactly fits the other. They are then tied firmly together with some soft twine and covered over with cowdung. In a short time (a month to six weeks) the two cut surfaces unite, when the branch may be cut away altogether from the tree and the top from the seedling, and the graft Mango can be taken away and planted.

2. The **Leaves** are long and narrow. *(a)* They are *so placed* on the stem *as to allow each to get its share of light:* the result is that the Mango tree gives a very dense shade.

(b) The *petioles* (leaf-stalks), besides placing the leaves in such a position that they can get light, also *save them from being torn* when the *wind* blows very strongly, as they are springy. If the leaves were fixed on more rigidly, the branches of the tree would get broken. As it is, the leaf sways from side to side and so escapes most of the wind-pressure. In a similar way, by bending down, they allow any rain-drops falling on them to drop to the ground from their tapering ends, instead of adding to the weight the branches have to bear. The leaf must, however,

have a certain rigidity in order to spread out a wide, green surface to the sun, and this is obtained by the system of veins or ribs.

(c) The leaves are also *"leathery,"* due to a thick epidermis (coat), which *reduces the rate of evaporation* of the water in the tree and is, therefore, of great importance to it. Every one knows that the Mango tree is evergreen and able to keep its leaves on all the year round. It thus preserves its roots much cooler than a tree which has at times no leaves, like the Teak. It is also able to go on storing up food all through the year; and as its fruit forms at a time when many other trees are leafless, this is another very great advantage to the Mango tree. But it does not follow from this that the Mango tree (or any other evergreen tree) has no resting period. "There is a definite periodicity—that is, a regular alternation of resting and working periods—noticeable in all tropical trees, although the seasons at which episodes, like the fall of leaves or the shooting of buds take place, vary greatly in different species, in different individuals of the same species, nay in different branches of one individual. In the Mango tree, for example, one or two branch systems may alone be putting forth the reddish-brown young leaves at a time when the rest of the crown retains the dark-green adult foliage" (J. M. F. Drummond).

(d) Further, the leaves are *smooth and shining,* somewhat like a looking glass. Now, we all know that light-rays are reflected from the surface of a mirror and not absorbed. The same is the case with the heat-rays which accompany the former, and by thus reflecting the heat-rays the temperature of the leaf is kept down and *a further source of evaporation taken away.* Evaporation increases in proportion as the temperature rises.

(e) The *young leaves* of the Mango tree are specially protected against any injury from excessive heat and light. When the tree is budding you can see all the young leaves *hang loosely* as if they were fading (pendent leaves, fig. 22). They have grown at an enormous rate and attained the size of full-grown leaves in a very short time. This being so, the tender cells of which they are composed and their contents could be easily destroyed by a vigorous

act of transpiration and by intense sunlight. The vertical placement of these leaves reduces the effect of heat and light. When the leaves are mature, the tissues are strained, and the leaves assume the ordinary horizontal position.

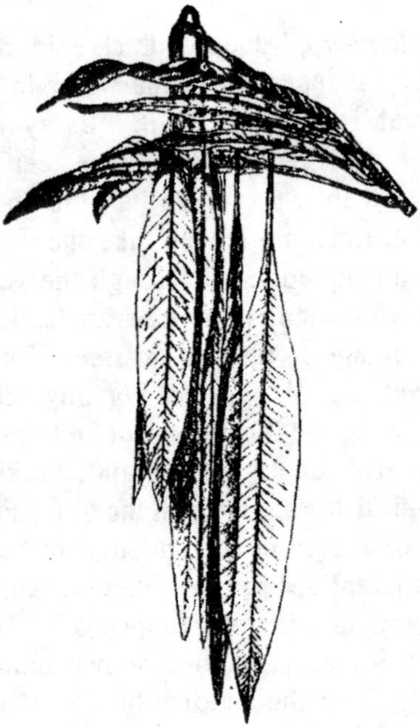

Fig. 22.—Young shoot of Mango with pendent leaves.

(*f*) The young leaves of the Mango tree are often *red* or *copper-coloured*. This is caused by a very active process of respiration or breathing, which, as we all know, is most active always in young organisms. For a plant, like an animal, breathes, that is to say, inhales oxygen in order to burn off some of its carbon, and thus produce the heat necessary for the various chemical processes carried on in its body to maintain its life and to grow. The gas produced by the combination of the inhaled oxygen and the carbon in the plant is carbonic acid gas, which is exhaled (see II. Part, Respiration). When the leaves become older, the process of breathing becomes slower and the copper-hue disappears from the leaves, changing into the ordinary green colour of leaves. The colour is due to little green granules, called chlorophyll* granules in the inner cells of the leaf. These granules have the power of feeding on the carbonic acid gas circulating through the air-spaces between the cells and to form starch under the influence of the sun's rays. This process is known as "assimilation" (see II. Part, Assimilation).

* From Greek *chloros*, green, and *phyllon*, a leaf.

(g) Another fact connected with the *arrangement of the leaves* may be noticed. They are so placed that, when rain falls, most of the water is carried from leaf to leaf *from the centre* of the tree *to its circumference* (compare the flow of water down a tiled roof), and it is on the outside of the tree that the young roots which alone can absorb water are most numerous. The big roots cannot do so, as they are covered with bark. This dripping tends to make the young roots grow outwards to where they can get water easiest. This again serves to give the tree a very wide root system, and prevents its being blown over by storms.

Fig. 23. — Flower of the Mango tree (considerably enlarged in size).

Fig. 24. – Branch of the Mango tree with fruits.

3. The **Flowers** are small and grow in erect panicles, which generally appear in January, February or March. The five sepals enclose five greenish-yellow petals which are alternate with them. There are five stamens, one of which only is perfect. The flower contains, in addition, a nectary which is an organ to secrete honey. You find it arranged round the ovary and consisting of five "fleshy" bodies. The existence of such a nectary suggests that the flower depends for fertilization on

attracting insects. The want of show in individual flowers is made up for by placing very large numbers of flowers close together.

4. When the **Fruit** begins to grow, the stalk of the panicle is not strong enough to hold it erect, so the fruit hangs down. Even so it would never be able to nourish all the fruits which might be expected from the number of flowers. Nature corrects herself. As the fruits grow, nourishment is gradually directed into from one to six of the most vigorous fruits, and the rest drop off gradually.

The ripe fruit is slightly compressed and is beaked, the point showing where the style was. It contains a "fleshy", palatable pulp under its leathery skin. The pulp surrounds a woody one-seeded nut with a fibrous beard. In the more inferior kinds these fibres run right through the pulp. The whole fruit is just a big drupe, like a peach or a plum.

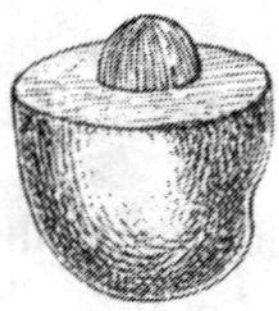

Fig. 25.—Transverse section of a Mango drupe with seed.

5. If the **Seed** is to grow into a new tree it must be carried away from the tree which bore it, as it could not thrive under the dense shade of the parent tree. This is provided for by the delicious pulp covering the seed, attracting men and animals to pick it up and carry it away. The seed is protected at the same time by a hard shell, formed by the inner part of the fruit cover (pericarp). We may notice a further instance of Nature's care in that the fruit, until it is ripe, is so acid that it cannot be eaten with any pleasure. Unripe seeds do not grow well.

Other Mangoes.

The Mango tree has some relatives affording useful fruits. One of these is the **Cashew Nut** (*Anacardium occidentale; Kan.* Gēru; *Mal.* Kaçumāvu; *Tam.* Mundirikai; *Tel.* Jīḍimāmiḍi; *San.* Çōphaharā; Plate 654), an American tree, brought to India by the Portuguese. What is generally called the fruit is the swollen flower-stalk or receptacle which bears the nut at the end. Both the juicy stalk and the nut are eaten. The former is very pretty, being coloured either a pale yellow or a brilliant red. The seed is protected by the cells in the cover being filled

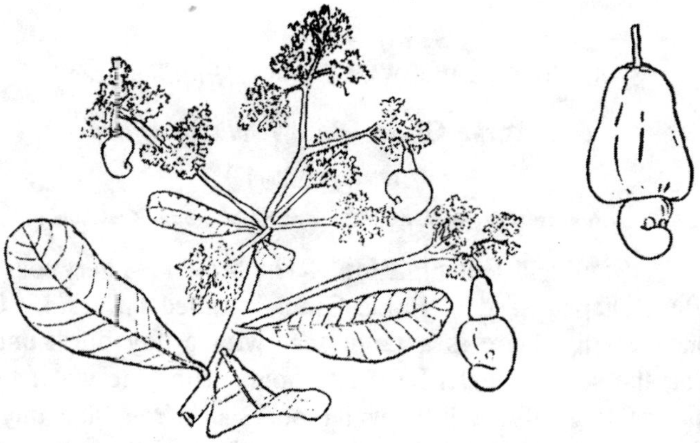

Fig. 26.— Branch of the Cashew Nut tree.

with an extremely acrid juice. They are, as a rule, roasted before being opened to get at the kernels which are edible. These nuts are exported in large quantities from the Malabar coast.

The **Indian Marking Nut** (*Semecarpus anacardium; Kan.* Gērkāyi, kēra; *Mal.* Čērmara; *Tam.* Çēngoṭaimaram; *Tel.* Jiḍičeḷṭu; *San.* Agnimukhi) yields a corrosive black juice, used by dhobies for marking clothes.

The fruit of the **Hog Plum Tree** (*Spondias mangifera; Kan.* Ambaṭa; *Mal.* Ambāḷam; *Tam.* Kāṭṭumā; *Tel.* Ambāḷamu; *San.* Āmrataka) is eaten, being a substitute for tamarind in curries. The tree flowers when it is leafless.

9. The Leguminosæ.

Trees, shrubs, or herbs. Leaves very often compound with stipules. Inflorescence racemose. Flowers zygomorphic. Sepals five, petals five. Stamens generally ten, free or variously combined. Ovary superior, of one carpel. Fruit a legume with the calyx attached. Seed exendospermous with thick cotyledons.

This large order is divided into three families:—

A. The Peaflower family *(Papilionaceæ).*

B. The Cassia Family *(Cæsalpinieæ).*

C. The Mimosa Family *(Mimoseæ).*

A. THE PEAFLOWER FAMILY *(Papilionaceæ).*

The Horse Gram Plant *(Dolichos biflorus).*

(Plate No. 650.)

(*Kan.* Huruḷi. *Mal.* Koḷḷu. *Tam.* Koḷḷu. *San.* Khalakula.)

1. **Seed** and **Germination**.—The brown leathery coat of the kidney-shaped seed of Horse Gram is called the *testa*. On its concave edge there is a large scar with a tiny pit at one end. Soak the seed in water for a few hours, wipe the water off and squeeze it gently: a little water oozes out from that tiny hole. The scar is called *hilum* and is the place of the attachment of the seed to the pod by means of a small cord called *funicle.* The testa can now be easily stripped off from the soaked seed and, if care be taken to notice the relative position of its outer markings and the inside structures of the seed, it will be found that the pointed end of a small stem is directed towards the tiny hole which is termed *micropyle.* The little stemlet, called *hypocotyl,* bears at its other end two thick lateral growths, the seed-leaves or *cotyledons,* and between them a tiny bud *(plumule)* of two folded, veined leaves. The seed, then, contains under its coat the future plant in miniature, the *embryo,* consisting of the hypocotyl, two cotyledons and the plumule.

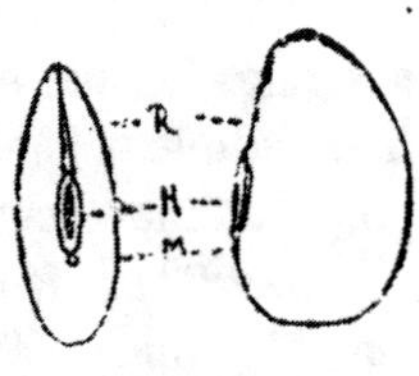

Fig. 27.—Bean seed.

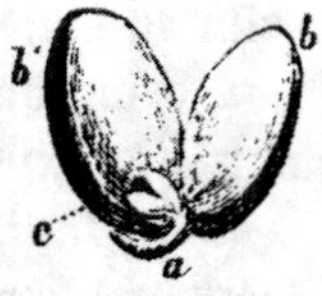

Fig. 28.—Seed of the Bean.
a. Hypocotyl. *b.* Seed-leaves. *c.* Plumule.

To observe the first stages of the growth of the plant, we allow the seeds to lie steeped in water for a longer time, say twenty-four hours. If we now examine them, we find the testa split across near the micropyle due to the expansion of the embryo and a tiny rootlet makes its appearance, which at once turns downwards, following the law of gravity. Now lay some of these germinating seeds in moist and loose earth and watch their gradual growth. The plantlet first establishes itself in the ground by growing a strong main root which sends off tiny branches all round. Meanwhile the hypocotyl also grows upwards, forming a hook with its two cotyledons still folded over the plumule. When the surface of the soil is reached, the hypocotyl becomes straight, and now the two cotyledons unfold. Soon after, the first pair of opposite, cordate foliage leaves are developed and all the parts above the soil grow green, the plantlet is now fully developed and consists of root, stem and leaves. As the stem continues growing vigorously, one leaf after another, now alternate and trifoliate, is developed and the cotyledons become thinner and finally fall off withered.

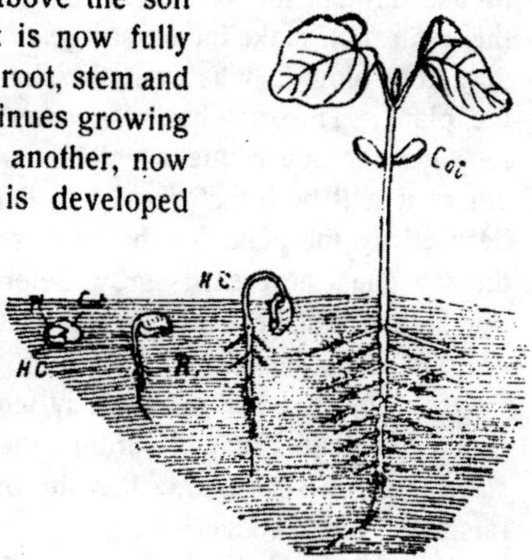

Fig. 29.—Germination of Horse Gram.

This process of germination gives us many things to learn.

(a) If beans or any other kind of seed are laid on a dry spot, they never germinate. They do so only when they are moistened. But then the question arises: why does the mother-plant not furnish the seeds with the water necessary for their germination at the very beginning, or why are the *seeds* always so *dry and hard* when they are produced in the fruit of the mother-plant? If it were not so, the seeds would try to grow as soon as they fall on the ground. But the tiny weak roots could not make their way into the hard ground, nor could they find any nourishment during the greater part of the year. For, at the time when plants generally ripen their seeds, the ground is dry and hard, the rains being over.

The seeds that had thus already begun to germinate would simply die of thirst, and it would be almost impossible for a plant to reproduce itself by seeds. In the case of annual herbs, such as Peas and Beans, etc., it would mean the extirpation of the genus in a very short time; for annual herbs are not able to live after flowering and fruiting. Another advantage of having hard, dry seeds is that animals, birds, and insects cannot destroy them so readily as they could if they were soft.

(b) We have noticed that *the first part* that comes out of the germinating seed *is the root.* There is a reason for this. The young plant must be fixed in the ground, when the hook-like stem breaks through the surface of it. The side roots, issuing from the main root, make the mooring so much the firmer. The wind may now blow in whatever direction it will, it cannot overturn the plant. The root has also other work to do. It must take in water and nourishment which is to be conveyed to the leaves where it will be transformed into that condition in which it can be used by the plant for the building up of new leaves. Now the root must necessarily grow before any other part can do so. For then it will be ready for its functions, and can at once send up food to the leaves when they are formed.

(c) The bud being extremely tender would suffer injury, if it had to force its way up through the soil. So the thick, strong *stem bends* in such a way that the bud remains below whilst it raises the earth above it.

(d) All the parts of the germinating seed are entirely colourless so long as they are within the coat of the seed and below the surface of the earth. They become green, however, when they come up to the light. It is the *action of sunlight which produces the green colour* in the stems and leaves of the plants.

(e) We know that the plants build up new parts from the food prepared in the leaves. But whence does the germinating seed take the materials necessary to form the root, the stem, and the leaves, as it has no leaves to prepare its food? We shall be in a position to give an answer to this question when we examine the seed-leaves carefully. These leaves are hard and full in the beginning, but gradually they become softer and softer

until they finally shrivel up and decay: the plant grew and *formed its root, its stem, and the first pair of leaves, all at the expense of the seed-leaves.* These were packed full with provisions such as the young plant would require at its first stage of growth. This food is, in the case of the Gram and many other plants, deposited in the seed-leaves themselves, whereas we find the food-store in a separate tissue, called endosperm, in the seeds of Rice and many other plants.

2. The **Stem** of the Horse Gram Plant is long and filiform and, therefore, cannot stand erect with its heavy load of leaves, flowers and pods. Many of these plants, as they are grown in the rice fields after the paddy is removed, do not find a support but trail along the ground. Their ends are raised and the tips of their stems are often slightly inclined and keep moving in a circular way as if they were seeking something to fasten on. When a support is found, their stems twine round it and the tips continue their circular movement. This movement is made in a direction contrary to that of the hands of a clock. There are other plants that move in the same direction as the clock's hands, *e. g. Dioscorea.*

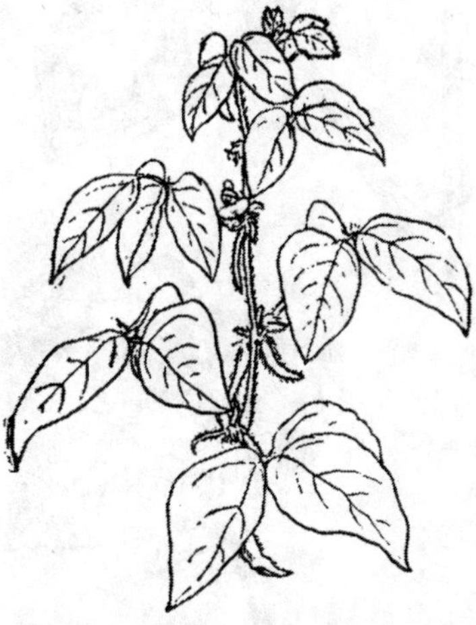

Fig. 30.—Stem of the Horse Gram.

3. **Leaves.**—*(a)* The first two foliage leaves are opposite and simple, as we have seen already. The succeeding ones are alternate and compound, consisting of a symmetrical leaflet at the end and two oblique leaflets on the sides. At the base of the common petiole as well as of each leaflet there is a pair of small appendages, called stipules.

(b) In the daytime the leaves are generally spread out horizontally; but when darkness sets in, the common stalk of the three leaflets begins to rise up, and the three leaflets descend

and hang down vertically (fig. 31, II.). We say, the leaf "sleeps" now. On the following morning it resumes its original position.

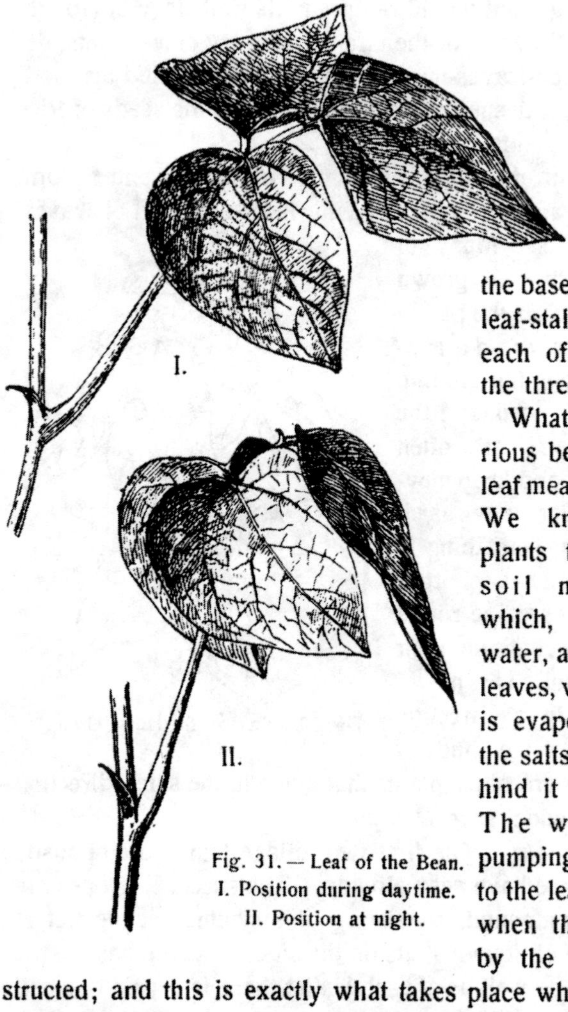

Fig. 31. — Leaf of the Bean. I. Position during daytime. II. Position at night.

These movements, which are very regular, are effected by the swollen joint called *pulvinus,* which can be observed at the base of the common leaf-stalk, as also on each of the stalks of the three leaflets.

What does this curious behaviour of the leaf mean for the plant? We know that the plants take from the soil nourishment which, dissolved in water, ascends to their leaves, where the water is evaporated leaving the salts of the soil behind it in the leaves. The work of thus pumping up new food to the leaves must stop when the evaporation by the leaves is obstructed; and this is exactly what takes place when dew settles upon the leaves. Now, it is a known fact that articles laid horizontally on the ground have a greater deposit of dew than such as hang vertically. It is therefore advantageous for the leaves to assume a position which prevents the dew from covering them and thus obstructing the process of evaporation which is so essential for their growth.

(c) When the heat of the sun is great so as to cause more water to be evaporated than the roots can absorb, the leaves are also seen to assume their "sleeping" attitude. As they place themselves parallel to the sun's rays, they are struck only by a few of them and at oblique angles (fig. 33); consequently they do not get so hot as would be the case, if they were at right angles to the sun's rays. This is another remarkable protective arrangement.

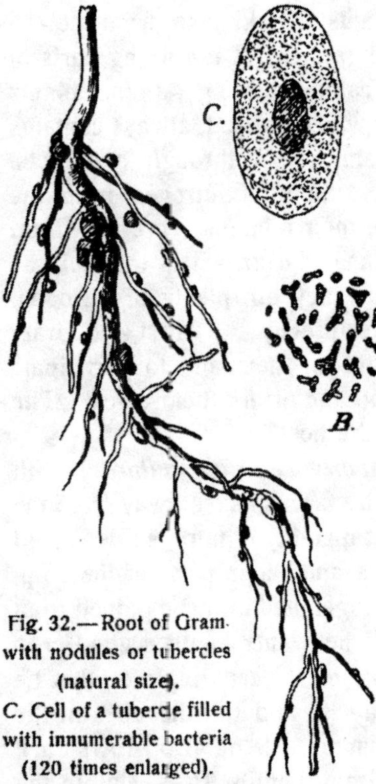

Fig. 32.—Root of Gram with nodules or tubercles (natural size).
C. Cell of a tubercle filled with innumerable bacteria (120 times enlarged).
B. Bacteria (800 times enlarged).

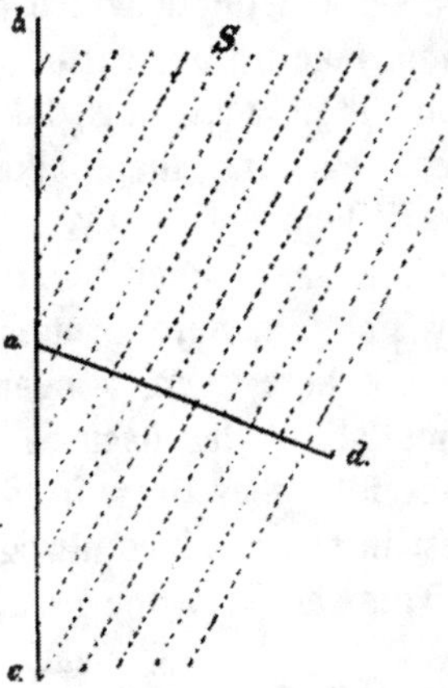

Fig. 33.—Solar rays falling vertically on *a d*, but slanting on *a c*, which are both of the same length.

4. **The Gram plant gathering Nitrogen from the Air.**—If you carefully pull a Gram plant out of the soil, you will notice numerous *nodules on the roots* (fig. 32) which are not accidental, but have their own little functions in the great household of Nature.

In each little grain of earth there are numerous minute germs, called bacteria. Some of these have the peculiarity of settling on the tiny root-ends of the plants of this order and grow as parasites on them, being nourished by their juices. They cause those parts of the root on which they settle to grow exuberantly, thus forming little nodules or tubercles (fig. 32, *C* and *B*) on the roots. Now, these bacteria, which are themselves tiny plants, have the power, unlike other plants, of taking in, from the air, nitrogen which is an essential ingredient of the living parts of plants and without which plants cannot thrive. As other plants cannot take nitrogen from the air which, indeed, always contains plenty of it, they must take it from the soil through their roots. These bacteria, then, take their supply of nitrogen from the air. When, eventually, they die, their remains serve the Gram plant, or any other plants that may be grown in that soil, as very good manure containing, as they do, plenty of nitrogen. We see here, then, a beautiful reciprocity. At first, the Gram plants allow the bacteria to settle on them and to participate in the food they draw from the soil and air for themselves. The guest in return gives nitrogen to the host.

This fact is of the *greatest importance to agriculture.* With each crop the farmer takes away from the field a great quantity of nitrogen deposited in the seeds and other parts of the crop, and this nitrogen had all been gained from the soil. If he wants another good crop next year, he must needs replace what he has taken away, and this he does in the shape of manure. If he also grows such plants as Gram, Lentils, Peas, etc., in the cropped fields, these themselves will help to manure the soil by acting as hosts to the bacteria which absorb nitrogen from the air.

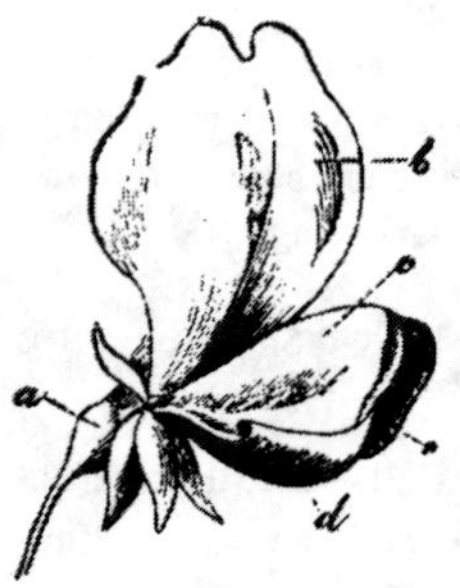

Fig. 34.—Butterfly-flower of the Pea.
a. Calyx. *b.* Standard. *c.* Wings. *d.* Keel.

5. The **Flower** has some **resemblance to a butterfly** (fig. 34). It is irregular, but a vertical section divides it into symmetrical halves. We call such flowers bilaterally symmetrical or zygomorphic (from Greek *zygon,* yoke, and *morphe,* shape).

(a) The *corolla,* supported by a cup-like calyx with five long points, consists of five petals which are of a creamy white colour and which differ from one another in shape and size. The largest of them is erect and is called the *standard* (fig. 34, *b*), because it stands up above the rest and shows its colours so boldly. It might also be called "sail", for it answers the purpose of one. The wind blows it round, so that it always turns its back to bad weather, and serves as a shield to the delicate parts within. The two lateral petals are called *wings* (fig. 34, *c*) and the two lower ones are so combined that they form a boat and are called the *keel.* These boat-like leaves enclose the stamens and the pistil (fig. 36) which latter consists of a long and hairy ovary and a style. In bud the petals are folded in such a way that the keel is innermost, the two wings fold over the keel, and the standard envelops all (imbricate æstivation).

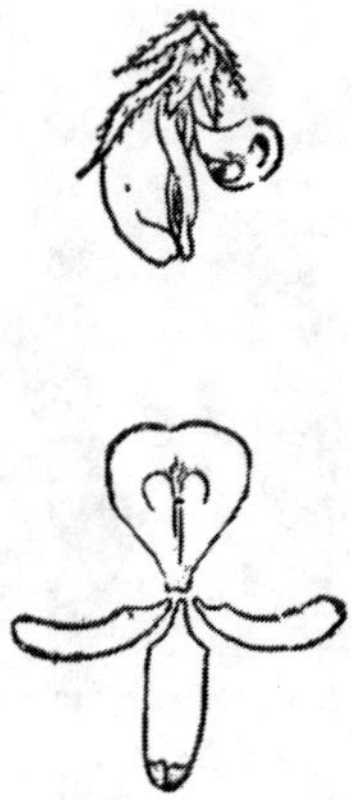

Fig. 35.— Flower of Horse Gram.

(b) There are ten *stamens.* The filaments of nine of them are united to a tube which, however, is not joined at the top. The filament of the tenth stamen lies in the slit. It is only through this slit that insects, attracted by the large standard and also by the sweet scent the flower exhales, can get at the honey which lies hidden within the tube of the filaments. While trying to get at the inner parts of the flower they must press down the keel. At once the style protrudes and touches the body of the visitor, which probably has come from another flower where its hairy body was covered with pollen, and the ovules are thus pollinated. At the same time, by the sudden jerk of the style, the stamens are

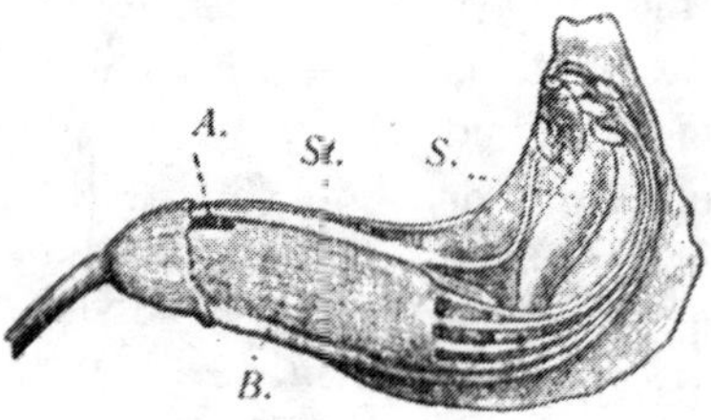

Fig. 36.— Keel of Pea flower (3 times enlarged). *B.* Bundle of the 9 combined stamens (only 4 of the 9 are visible). *St.* Single stamen. *A.* Access to honey. *S.* Style with stigma and brush.

caused to spill their pollen on the insect which now with a new load of pollen flies to another flower. Now, as the wings and the keel of the flower are, by a sort of joint at their base attached to one another very closely and firmly, it requires considerable strength to press them down so as to get at the honey in the interior. Not all insects have the strength necessary to do this. Bees are strong enough to overcome all the difficulties on their way to the honey, and it is chiefly by them that these flowers are fertilized. Some of them, it is true, try to get at the honey by a shorter way. They break in like thieves and bite a hole through the flower-leaves at the base.

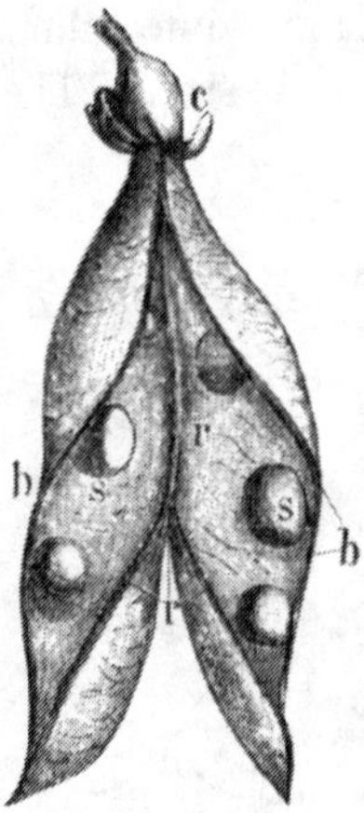

Fig. 37.—Legume of the Pea.
c. Calyx. *s.* Seeds.
r. Midrib.
b. Ventral suture.

6. The **Fruit** (fig. 37) is a pod or *legume*, consisting of a single carpellary leaf, which is folded inwards in its midrib having the edges seamed together (ventral suture). The pod contains several seeds in one row, all attached alternately to the two edges of the seam of the fruit-leaf which form the concave edge of the legume. The calyx remains attached to the fruit and is, therefore, called persistent. The point of the pod, more or less withered, is, of course, the style of the flower. When ripe, the legume splits both at the midrib and at the seam, thus dividing into two halves. There are various modifications of this typical legume in other genera of the Leguminosæ. The legume of *Pongamia* is indehiscent (not splitting) and one-seeded; that of *Smithia* and *Desmodium* jointed and when ripe dividing into its separate joints; *Butea* and *Dalbergia* have winged legumes; and *Crotalaria* inflated ones, etc.

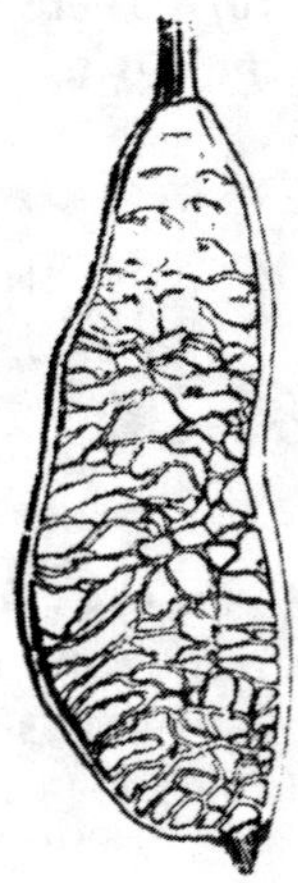

Fig. 38.—Legume of *Butea frondosa*.

Other Papilionaceæ or Butterfly-flower Plants.

This tribe is a very large one. We find papilionaceous plants cultivated in the fields for the fibre of the stem like the **Sunn Hemp** (*Crotalaria juncea; Kan.* Saṇabu; *Mal.* Čaṇaka; *Tam.* Čaṇal); for their nutritious seeds like the **Bean** (*Phaseolus trilobus; Kan.* Āvare; *Mal.* Čeruviḍukōl; *Tam.* Elippayaru), the **Pea** (*Pisum sativum*; *Kan.* Baṭāṇi), the **Bengal Gram** (*Cicer arietinum; Kan.* Kaḍale; *Mal.* Kaṭala), the **Green Gram** (*Phaseolus mungo; Kan.* Uddu; *Mal.* Uḷunnu; *Tam.* Uḷundu), the **Alasandi Bean** (*Vigna catjang; Kan.* Alasandi; *Mal.* Kōṭṭapayaru; *Tam.* Kārāmaṇi), the **Ground Nut** (*Arachis hypogœa; Kan.* Nelakaḍale; *Mal.* Nelakaḍalaka; *Tam.* Vērkaḍalai); for the dye obtained from the leaves, like the **Indigo Plant** (*Indigofera tinctoria; Kan.* Nīli; *Mal.* Avari—Plate No. 634). But we can also see them wild in the forests, as for instance the **Rosewood** (*Dalbergia latifolia; Kan.* Bīṭi; *Mal.* Viṭṭi; *Tam.* Iṭṭi), which yields an excellent timber, or the **Coral Tree** (*Erythrina indica*; *Kan.* Hoṅgara; *Mal* Muḷḷumurika), with its beautiful shining red flowers, and so many others; while the creepers **Wild Liquorice** (*Abrus precatorius; Kan.* Gurguñji; *Mal.* Guñja; *Tam.* Kunri), and **Shankapushpa** (*Clitoria ternatea; Kan., Mal.* Śaṅkhapushpa) adorn our hedges with the pretty scarlet seeds of the former and the large blue and white flowers of the latter.

Fig. 39.—Flower of the Tamarind tree.

B. THE CASSIA FAMILY (*Cæsalpiniaceæ*).

The **Tamarind Tree** (*Tamarindus indicus; Kan.* Huṇise—Plate No. 651) is a stately tree with a round crown, cultivated for its fruit. Leaves evenly pinnate, sensitive. In young leaves large reddish stipules which become scarious after a time and are shed. Leaflets ten to fifteen pair, oblique, oblong, entire, emarginate at the apex, glabrous. Flowers in lax racemes.

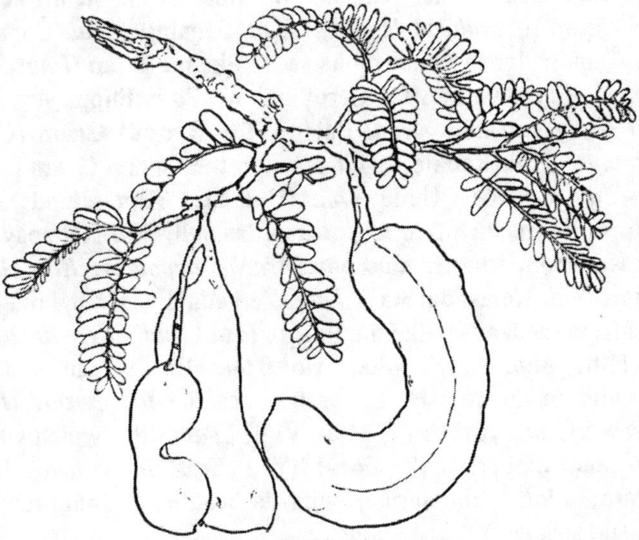

Fig. 40.—Branch of the Tamarind tree with fruits.

Bracts deciduous. Receptacle hollowed. Sepals four. Petals three. Stamens three, connate at their base and showing marks of six or seven rudimentary stamens. Pod with a brittle exocarp, a pulpy sour mesocarp and each seed enveloped in a hard endocarp.

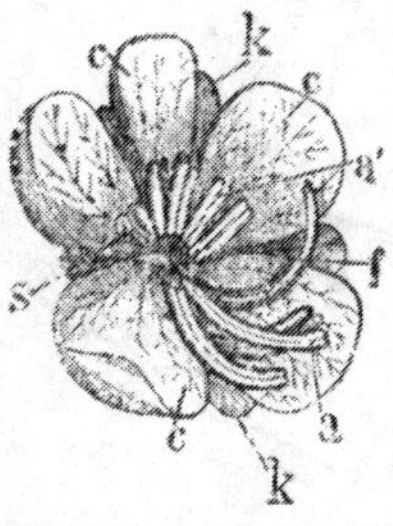

Fig. 41.—Flower of *Cassia*. *k*. Calyx. *c*. Corolla. *a*. Long Stamens. *a'*. The shorter stamens. *s*. Sterile stamens. *f*. Pistil.

The **Indian Laburnum** (*Cassia fistula; Kan*. Konde; *Mal*. Konne; *Tam*. Kovrai; *Tel*. Rēlačeṭṭu; *San*. Suvarṇaka) is one of the most beautiful jungle trees when in full flower. The fragrant, golden flowers hanging down in long, drooping racemes appear after the first rains together with the pinnate leaves which the tree sheds in the cold season. Sepals and petals each five. Petals almost equal, imbricate in bud and rested on the mouth of the hollow receptacle. Stamens ten of various length. The three lower ones have long and curved filaments, their anthers opening by longitudinal slits, the next four stamens have short filaments and anthers opening by basal pores, the rest have minute anthers without pollen. Fruit a long cylindrical indehiscent legume.

Other common Cæsalpiniaceæ

are the **Gold Mohur** tree (*Poinciana regia: Kan.* Doḍḍa Ratna-

Fig. 42. — An "umbrella tree": *Poinciana regia.*

gandhi: fig. 42), the **Peacock's Crest** (*Cæsalpinia pulcherrima; Kan.* Ratnagandhi), both of which have ten equal stamens, arranged in two whorls. One of the petals, the standard, is unlike the other four. Æstivation imbricate as in the Papilionaceæ, but the standard innermost in bud.—Of peculiar interest are the two-lobed leaves of **Bauhinia** (*Kan.* Mandāra), which at night fold over their midribs.

C. THE MIMOSA FAMILY *(Mimoseæ).*

The Babul Acacia *(Acacia arabica).*

(Plate No. 628.)

(*Kan.* Karijāli. *Mal.* Karuvēlam. *Tam.* Karuvēl. *Tel.* Nallatumma. *San.* Barbūrah.)

This well-known tree is essentially a tree of dry regions and is, therefore, be met with more commonly in the interior of India than on the sea-coast. To such places it is ingeniously adapted.

1. **Leaves,** with large blades, would allow too active an evaporation for the thirsty places the Acacia lives in. Their surface

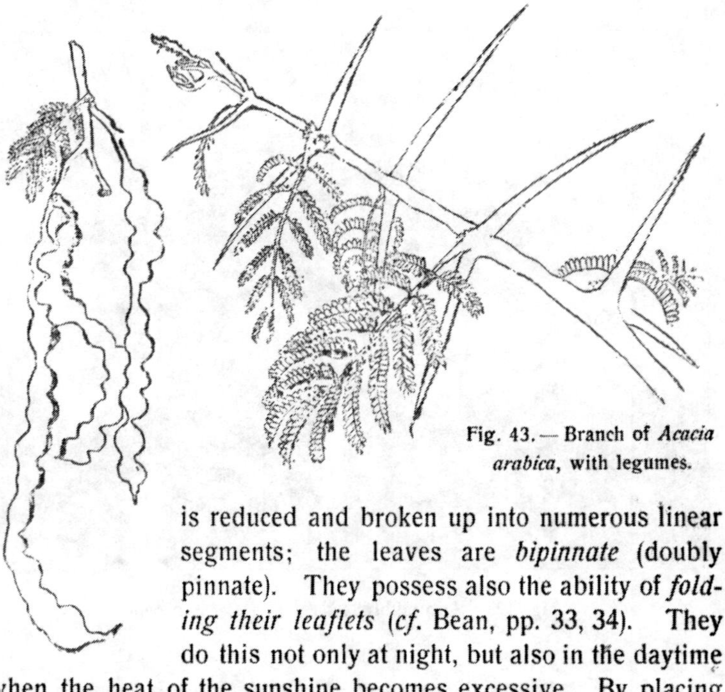

Fig. 43.— Branch of *Acacia arabica*, with legumes.

is reduced and broken up into numerous linear segments; the leaves are *bipinnate* (doubly pinnate). They possess also the ability of *folding their leaflets* (*cf.* Bean, pp. 33, 34). They do this not only at night, but also in the daytime when the heat of the sunshine becomes excessive. By placing the leaflets vertically they cause the sun's rays to fall on them in acute angles and so reduce the heating effect of the sunshine and, thereby, check the action of evaporation through their surface.

2. The **Stipules** at the base of each leaf are transformed into long, straight *thorns* which the shrub can employ effectively as weapons against animals that would otherwise feed on them.

Protection is also afforded by an *astringent acid,* called tannin, contained *in the bark.* If the bark is damaged, gum trickles out of it and covers the wound. As the Acacia tree is one of the few plants that grow in deserts, it can make good use of such means of defence.

3. The **Flowers** (fig. 44) are small, but are grouped in round heads. As the tree flowers in desert regions and at such a time when there is no rain, it can dispense with the many arrangements by which, in other plants, the pollen of the stamens is protected against bad weather. So the floral envelopes (calyx and corolla) are considerably reduced (see fig. 44, *c* and *k*), and the numerous stamens protrude widely from them. In fact, the flower-heads look yellow from the pollen only.

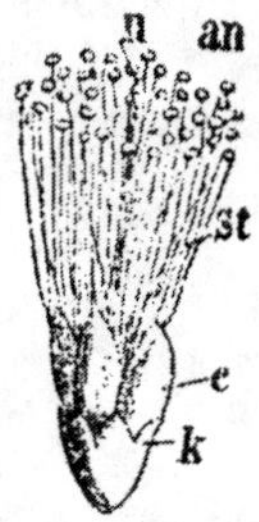

Fig. 44.—Flower of the Acacia. *k.* Calyx. *c.* Corolla. *st.* Stamens. *an.* Anthers. *n.* Stigma of style. (Much enlarged.)

4. The **Legumes** (Plate No. 628, 4) are flat and narrow and depressed between the seeds.

5. **Uses for man.**—The tree is useful in many ways. The wood is very hard and can be employed for all purposes for which a hard wood is required, such as ploughshares, knees and ribs of country boats, naves of wheels, etc. The bark is employed in tanning. The pods form a valuable food for cattle, and the young branches are a favourite food of camels and goats. The bark yields also gum, which is an article of general commerce.

Other Mimoseæ

are the **Red-wood Tree** (*Adenanthera pavonina; Kan.* Mañjeṭṭi, Mañjāḍi) known for its scarlet seeds; the **Soapnut Acacia** (*Acacia concinna; Kan.* Sīge); the **Fragrant Acacia** (*Acacia Farnesiana; Kan.* Pīkjāli; *Mal.* Karivēlam); the **Rain Tree** *(Pithecolobium saman)*; the **Korukapuli** (*P. dulce; Kan.* Čakli; *Tam.* Koḍukkapuḷi); the **Sirissa Tree** (*Albizzia Lebbek; Kan.* Bāge; *Tam.* Vāgai), and **Entada** *scandens* (*Kan.* Hallekāyi), the latter a liana with legumes of immense size. Another well-known mimosa is the **Sensitive Plant** (*Mimosa pudica; Kan.* Nāčikegiḍa) so called from its highly sensitive leaves which fold and bend when touched (fig. 45). "The Sensitive Plants derive protective advantages from these movements. They often cover large tracts of land, and grazing animals may be often attracted

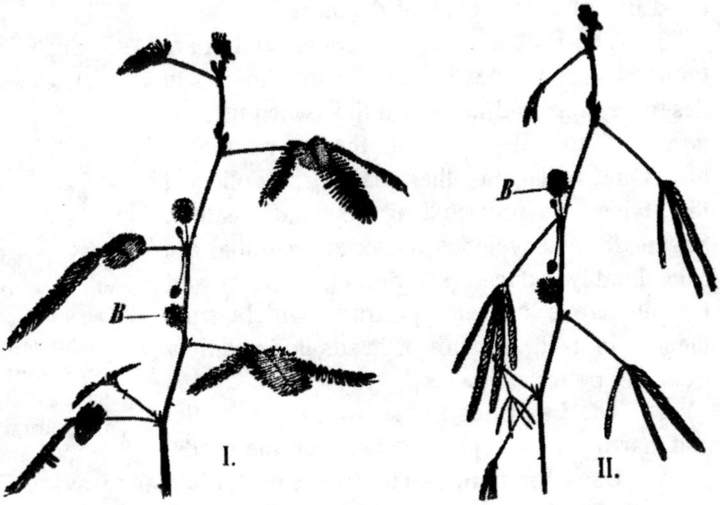

Fig. 45.—*Mimosa pudica*. I. Position at day, or when circumstances are favourable. II. Position at night, or when touched.

towards them by their bright green foliage. But what happens? The very first plants the animal approaches droop their tempting leaves, sensitive even to the vibration of the ground caused by the invader; and should it step in amongst them, the tempting and juicy foliage recedes before it, for one plant conveys the shock to its neighbours by the touch of its own leaves as they drop. Thus, what was a moment before a mass of tempting green leaves becomes almost instantly in appearance very scrubby fare for the animal, whose appetite expected much better refreshment. The stems are protected with strong and sharp spines. And now, when they have turned down their leaves out of harm's way, they present to their enemy for its first nibble nothing but prickly stems, so that should the intruder not be awed by their uncanny movements, but proceed with its intention, its first mouthful would scarcely be agreeable after its richer anticipations."

Various trees of this group form the characteristic flora of desert regions. Some of them have a peculiar mode of growth. Their *crown* is not round like that of the Mango tree, but *flat like a palm-leaf umbrella* (čatra, fig. 42). It is thus, indeed, more exposed to the sun's rays, but it presents its edges to the scorching winds instead of a large surface and thereby is better protected from loss of moisture by transpiration. The effect of the sun's rays is lessened by the trees' power of folding their leaves.

10. The Rose Family

(Rosaceæ).

Trees, shrubs, or herbs. Leaves alternate, stipulate. Flowers radial. Sepals and petals each five, imbricate in bud. Stamens numerous, free, incurved in bud. Carpels distinct (apocarpous). Fruit often fleshy, and in its formation aided by the receptacle.

The Rose *(Rosa centifolia).*

(*Kan.* Gulābi. *Mal.* Paninīrpu.)

The Rose is the queen of flowers. Its graceful shape, beautiful colour, and delicious scent make its charm irresistible. It is the symbol of youth, of innocence and of beauty. With roses we decorate our houses on festive occasions, and make beautiful the graves of our beloved ones.

1. The **Double Rose** *(Rosa centifolia)* is cultivated in our gardens. If a bush with double flowers is not attended to and pruned and manured, it will in course of time yield single flowers. One of the species with single flowers is the Hedge Rose *(Rosa canina)* which in colder climates grows wild in hedges.

2. The *prickles* on the **Stem** are worthy of special notice. They are sharp and bent down (recurved). With their help the weak stems seek support on other plants that are stronger than themselves. But the prickles are also powerful weapons against enemies, such as cattle, which would feed on them, or snails which would crawl up to the tender leaves, or mice which would eat the sweet fruits, called hips. It may be noticed that the prickles of the Rose are like hairs only loosely

fixed on the bark. They are growths of the skin (epidermis) and differ in this respect from the spines of the Acacia, which are modified stipules, and also from the axillary spines of the Lemon tree, which are modified branchlets.

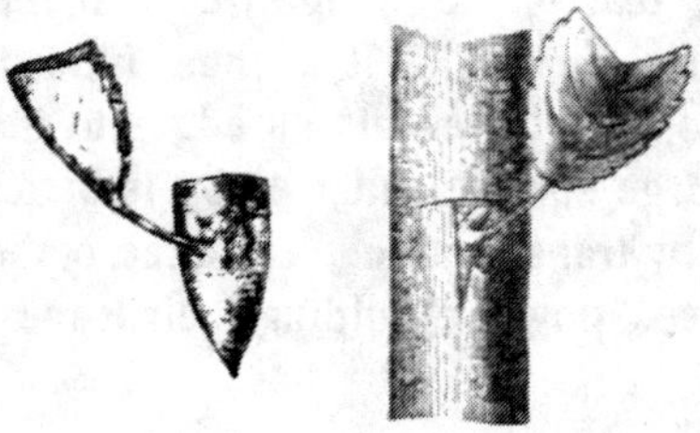

Fig. 46.— Grafting by budding.

As in the Mango tree, the finer sorts of Roses are *grafted* on the wild sorts by budding, *i. e.*, by inserting the bud of a superior kind under the bark of the inferior (fig. 46).

3. The Rose **Leaves** consist of a long middle rib with five or seven leaflets, of which two are always opposite, the midrib ending in a single leaf. Such leaves are called *imparipinnate,* that is, unevenly pinnate (compare it with a feather). At the base of the leaf-stalk are two *stipules* which in young branches embrace a younger leaf; between the stipules of this, the next younger again is covered, and so on. In this manner the inner, very tender leaves are covered by the outer ones by their sheathing stipules. The young leaflets are folded and laid together like the leaves of a book. It may also be noticed that the young leaves of some sorts of Roses are red coloured. (See Mango, page 26.)

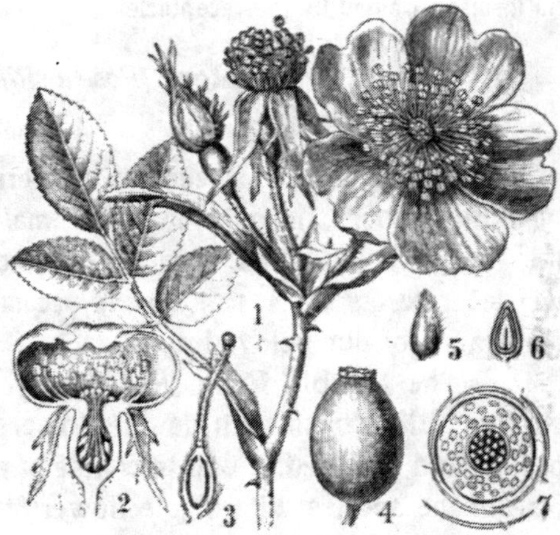

Fig. 47.— Hedge Rose *(Rosa canina).*
1. Flowering branch. 2. Longitudinal section of bud. 3. Ovule with style. 4. Hip (fruit). 5. and 6. Seed. 7. Diagram.

4. In the **Flowers** of the Hedge Rose (fig. 47) we can distinguish first a green cup-like receptacle, containing numerous distinct pistils and crowned by five sepals, five petals, and numerous stamens. Sepals and petals are imbricate in bud, and the stamens bent inwards (incurved). The cultivated Rose has, however, numerous petals, which are formed by the transformation of some of the stamens. This is sometimes easily seen as some of the petals in the middle occasionally bear pollen bags.

5. The **Carpels** are seated disunited or free in the hollowed-out "fleshy" top of the flower-stalk, which becomes a beautiful red colour when ripe, and is made up of a soft sweet pulp in order to attract birds by whose means the seeds can be scattered

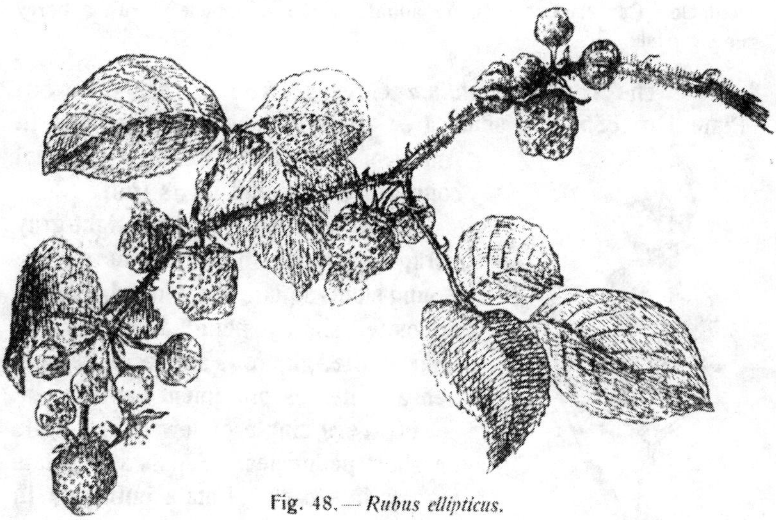

Fig. 48.—*Rubus ellipticus.*

far and wide. To prevent the seeds being digested by the birds the pericarps are thick-walled and are provided with prickly hairs and so are not damaged even if swallowed, which is rather unlikely, as the birds do not like such hairs.

6. The **Scent** of the Rose is derived from a volatile oil which evaporates easily and leaves no greasy stain, if applied to paper. This oil is extracted from the petals, by distillation, and then sold as a precious perfume, known as Attar of Roses.

The Rose family *(Rosaceæ)* is but little represented on the plains of India. Many fruits, such as the Apple, the Pear, the Almond, the Peach, the Cherry, the Plum, Strawberries and Raspberries belong to this family, but come to perfection only in cooler climates. *Rubus ellipticus* (fig. 48), *R. moluccanus,* and *R. lasiocarpus* are very common on the hills.

11. The Myrtle Family

(Myrtaceæ).

Trees or shrubs. Leaves mostly opposite, simple, entire, often with glandular dots. Flowers radial. Sepals and petals four (or five) each, inserted with the numerous stamens on the mouth of the hollow receptacle. Carpels two or more adnate to the receptacle. Fruit a berry or a capsule.

1. The **Guava** *(Psidium Guyava; Kan.* Pēraḷe—fig. 50; Plate No. 655). A small tree or large shrub, indigenous in America, cultivated in most tropical countries on account of its fruit.

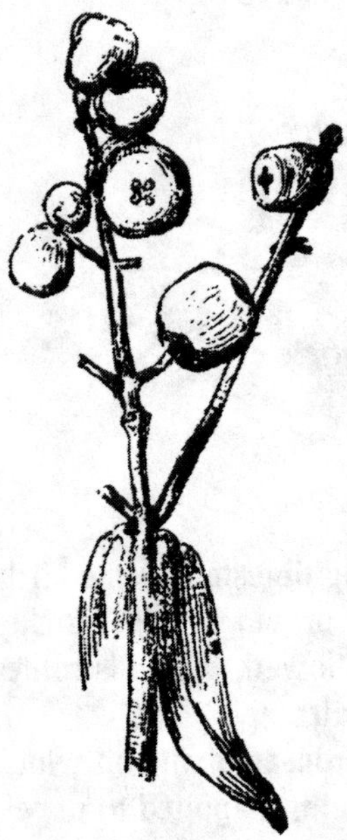

Fig. 49.—*Eugenia zeylanica.*

Stem with smooth and light-gray bark, which is shed in thin sheets. Young stems square and winged. Leaves opposite, shortly petioled, elliptic-oblong, entire, glabrous above, pubescent beneath, nerves prominent beneath.

Flowers single or few in leaf-axils on short peduncles. Sepals and petals four or five each. Petals imbricate in bud and deciduous. Stamens numerous, bent inwards in bud, and inserted on the mouth of the hollow receptacle. Ovary inferior, enclosed in the receptacle, two or more-celled with numerous seeds. Fruit a berry, crowned by the persistent calyx.

2. The **Jamoon** *(Eugenia jambolana; Kan.* Nēraḷe; *Tam.* Nāval; *San.*

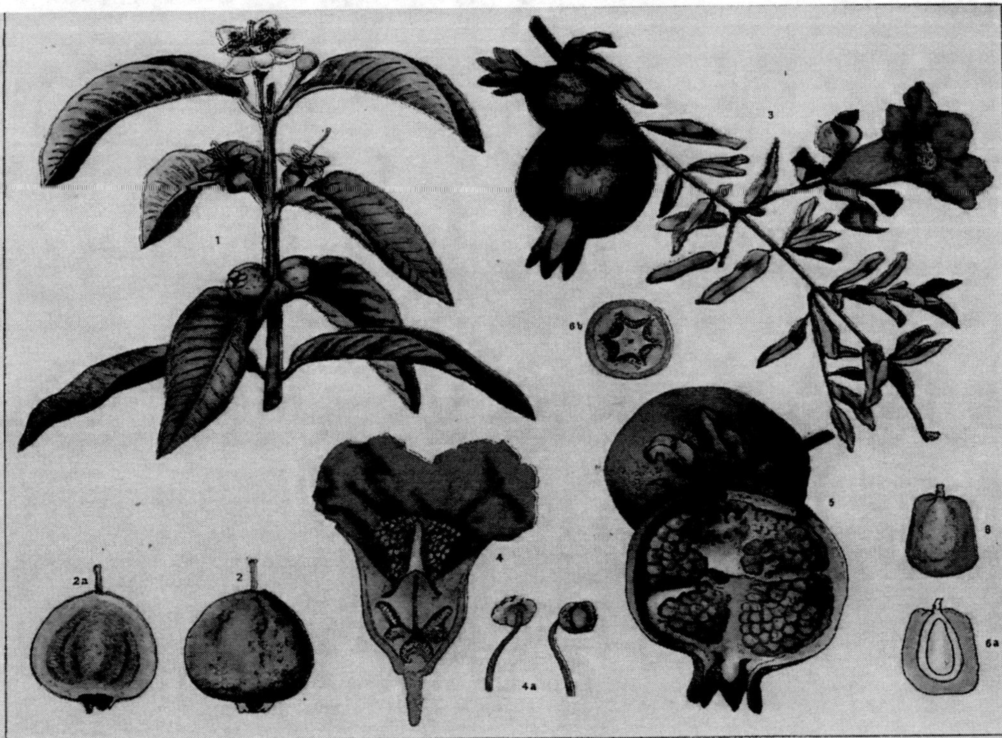

Fig. 50.—GUAVA (*Psidium Guyava*).—POMEGRANATES (*Punica Granatum*).

Jāmbavam), a large tree with light gray bark. Leaves elliptic-lanceolate, acuminate, coriaceous; secondary nerves running closely parallel and meeting in one just inside of, and parallel to the margin. When held up to the light small transparent dots are seen: these are marks of a volatile oil contained in glands: when the leaves are bruised, the scent of that oil will be noticed. Flowers in their morphological structure are like those of the Guava, the deciduous petals and the slender filaments of the numerous stamens being particularly noticeable. The latter make the flowers showy (compare the flowers of Acacia, page 42). Inflorescence cymose on the previous year's wood. Fruit a one-seeded berry, edible.

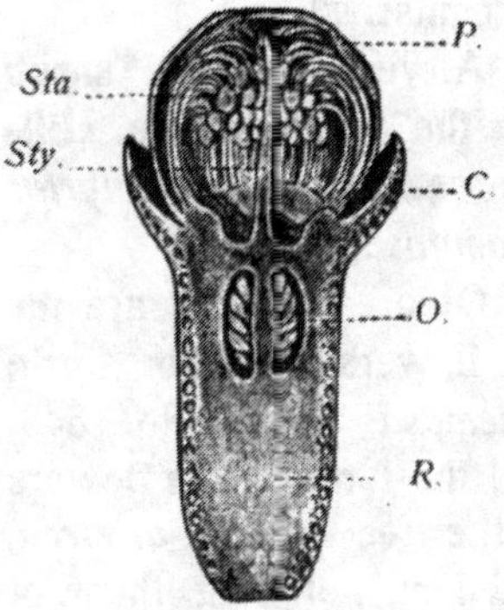

Fig. 51.—Flower-bud of the Clove *(Eugenia caryophyllata)*. *R.* Receptacle. *O.* Ovary. *C.* Calyx. *P.* Petals. *Sta.* Stamens. *Sty.* Style. The dots below the epidermis denote small cavities containing aromatic resin.

3. Several species of the genus Eugenia are found on the hills; the commonest in the plains are *E. corymbosa* (*Kan.* Kuṇṭāla; *Mal.* Nēral), *E. zeylanica* (*Kan.* Guḍḍapannēraḷu, fig. 49) both yielding a very good fuel-wood. The fruit of *E. malaccensis* (Rose-apple) is eaten, and from the *E. caryophyllata* we obtain the well-known spice, Cloves (fig. 51).

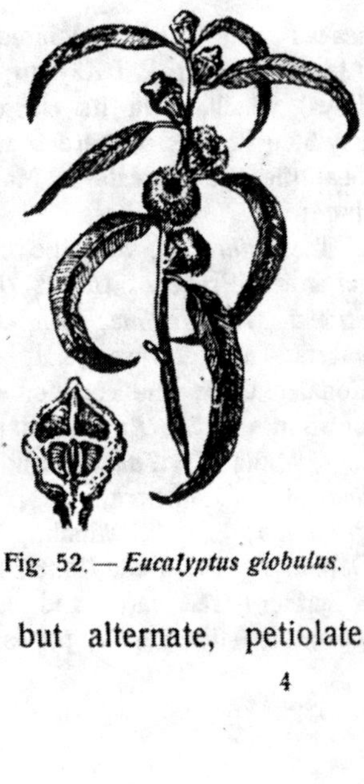

Fig. 52.—*Eucalyptus globulus.*

4. The **Blue Gum Tree** *(Eucalyptus globulus)* and other species of the genus Eucalyptus, indigenous in Australia, are now cultivated in India, chiefly on the hills, on account of their rapid growth, producing a great quantity of fuel. The leaves are opposite, sessile, cordate, glaucous-gray, and horizontal in saplings, but alternate, petiolate,

lanceolate-falcate, green, and vertical in the adult trees. The fruit is a hard, leathery capsule opening by a lid which falls off (circumsciss).

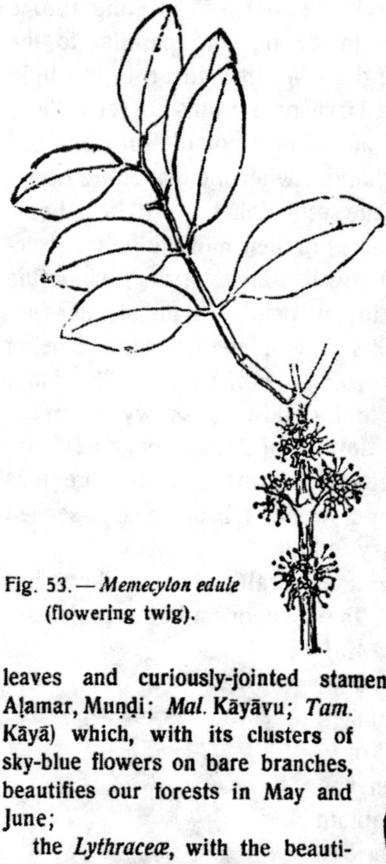

Fig. 53.—*Memecylon edule* (flowering twig).

A very common shrub on the hills is the **Hill-Gooseberry** *(Rhodomyrtus tomentosa)*.

Of great beauty are the red flowers of **Barringtonia** *racemosa* (Plate No. 643) and the large, white flowers of the tree **Careya** *arborea*, and last, not least, those of the **Bridal Myrtle** *(Myrtus communis)*.

Allied Families

are the *Melastomaceæ*, of which we mention two handsome shrubs, *viz.*, *Melastoma malabathricum* (*Kan.* Doḍḍa-nekkare; *Mal.* Kaḍaḷi) with three-nerved leaves and curiously-jointed stamens; and *Memecylon edule* (*Kan.* Aḷamar, Munḍi; *Mal.* Kāyāvu; *Tam.* Kāyā) which, with its clusters of sky-blue flowers on bare branches, beautifies our forests in May and June;

the *Lythraceæ*, with the beautiful tree, **'Pride of India'**, *(Lagerstrœmia flos reginæ; Kan.* Maruvāčala; *Mal.* Niruvenḍēku; *Tam.* Kodalemukki); the common shrub **Lawsonia** *alba (Kan.* Madarangi; *Mal.* Mailāñči; *Tam.* Marudōnri): and the **Pomegranate** *(Punica granatum; Kan.* Dālimbe; *San.* Dāḷika: fig. 50; Plate No. 655) with a leathery, red calyx and usually more than five sepals, petals, and

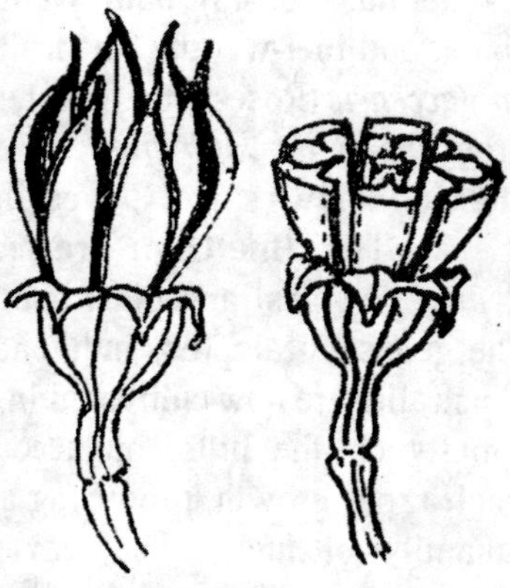

Fig. 54.—*Lagerstrœmia flos reginæ* (capsule).

carpels each. Petals, like those of Lagerstrœmia, crumpled in bud. Carpels arranged in two tiers, inside the hollow receptacle. Fruit a berry with leathery pericarp and numerous seeds embedded in a "fleshy", sweet, pellucid testa;

the *Combretaceæ*, — with the well-known **Rangoon Creeper** *(Quisqualis indica)*, and several useful timber-trees of the genus **Terminalia,** *e. g.*, *T. tomentosa (Kan.* Banapu; *Tam.* Karimardu) with five brown coriaceous wings in the one-seeded nut; and *T. paniculata* (*Kan.* Maruva; *Tam.* Pūmardu) with a one-winged nut.

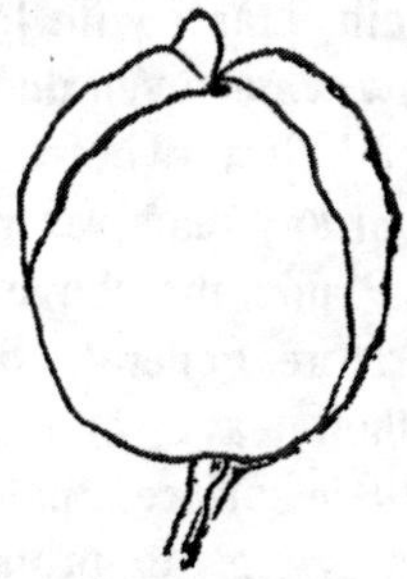

Fig. 55. — *Terminalia tomentosa* (winged fruit).

12. The Mangrove Family
(Rhizophoraceæ.)

Trees or shrubs, mostly in brackish water. Leaves opposite, mostly coriaceous, entire; stipules interpetiolar, deciduous. Flowers radial. Ovary inferior. Fruit one-seeded.

The coasts of India and other tropical countries are lined with peculiar flora called **'the Mangrove'** to which belong several trees of the *Rhizophoraceæ,* e. g. *Rhizophora mucronata* (*Kan.* Kāṇḍel), and *Bruguiera parviflora,* as well as some shrubs of other families, *viz., Avicennia officinalis* (*Kan.* Upruṅja), the spinous *Acanthus ilicifolius* (*Kan.* Hoḷečuḷḷi), and the fern *Acrostichum aureum.* The peculiar habitat of these plants has conditioned their peculiar structure.

1. **Xerophilous and hygrophilous habits of plants.**— Even a child knows that one of the conditions for the healthy growth of a plant is a regular supply of water. There is a constant flow of water in a tree from its root up to the leaves. But this flow must not be too weak, or else the leaves begin to fade and the plant dies. The flow must also not be too strong, or else the plant is overfed, becomes sickly and finally dies.

Now, the supply of water is not the same in every place. Plants growing near a river-side always have water enough to drink; plants growing on dry ground, however, have very little water to feed on. The supply of water is also not the same at all times of the year. There is plenty of water for every plant during the rainy season, but hardly any during the hot season. Again, plants with long and deep roots, such as large trees, can draw water even during the rainless season from springs hidden deep below, whereas many small plants with short roots cannot reach any such reservoir.

Plants that have a rich supply of water at their disposal, therefore, generally possess certain arrangements in the structure of their leaves and stems to facilitate the transpiration of water at their surface, that is to say, their structure is *hygrophilous** (see "Helps to promote Transpiration" in II. Part). But such plants as have only a scanty supply of water must be very frugal with it, they must reduce the process of transpiration to a minimum and their structure is said to be *xerophilous*** (see "Means to check too much Transpiration" in II. Part). Plants that are able to adapt themselves to various conditions as regards the supply of water in the different seasons are said to be *tropophilous*† (see Teak Tree).

2. **Xerophilous structure of the Mangrove.**—If we examine the leaves of a Mangrove tree, *e. g., Rhizophora mucronata,* under a microscope, we find that they are covered with a thick coat, called the cuticle, preventing the evaporation of water accumulated in the cells of the epidermis. We further find that there are a number of additional layers of cells filled with water below the epidermis of the upper side of the leaf which may be said to be a reservoir of water. The stomata are deeply set in the epidermis. Many of the interior cells contain a slimy substance. The leaves of the tree have a decidedly xerophilous structure like so many plants growing in dry places.

At first sight, this fact must cause some surprise to the observer; for the Mangrove tree lives on moist soil, and we should, therefore, expect a hygrophilous rather than a xerophilous structure. However, if we think about it a little more, we understand that the Mangrove must necessarily be thus equipped. The

---

* From Greek *hygros,* wet; and *philos,* loving. ** From Greek *xeros,* dry; and *philos,* loving. † From Greek *tropos,* a turn.

soil in which the tree grows is moist, but salt. If the tree were to absorb a large quantity of salt, the vessels in both root and stem would be clogged by salt, which would cause the death of the tree. The tree must, therefore, reduce the quantity of water taken in by the roots to a minimum, which is effected by diminishing the action of transpiration. The water absorbed is thus not freely transpired, but retained by the leaves, and the percentage of salt is kept at a low rate.

The diminution of the flow of the sap is required for a second reason. Breezes are constantly blowing on the sea-coast. (The heat of the sun's rays warms the land much more than the sea, then the hot air on the land rises, and the cool air from the sea rushes in to replace it. That is the ordinary sea-breeze by day. The reverse takes place at night. The land cools down much sooner than the sea, the warm air over the sea rises and is replaced by a stream of air from the land, which we call land-breeze.) A plant like the Mangrove tree that is thus constantly exposed to winds is likely to wither easily and, hence, must be protected against too great a loss of water by transpiration.

3. **Rooting in the muddy soil.**—The soil in which the Mangrove grows is loose and muddy, and the tides continually shake the plant. The plant has to meet these unfavourable circumstances and is for this purpose admirably provided. It sends out from its trunk numerous horizontal adventitious roots which by means of vertical shoots manage to get a firm hold in the soft mud. Such adventitious roots spring even from the branches. Mangrove swamps thus obtain a peculiar appearance, especially at low-tide, when the whole grove seems to be lifted over the water, as if standing on stilts, or looks like so many spiders standing on their long legs (fig. 56). The water can now flow to and fro under the trunks, just as under the piles of a pier, the stems remaining firm.

4. **Germination of the Seed.**—But what happens with the seed? Is it possible for the germinating seed to take root in such an unsteady soil as the beach affords? Provision is made also for this difficulty. Seeds, as a rule, take a long rest before they germinate. The seed of the Mangrove tree, however, is an exception

Fig. 56. — Mangrove growing on the banks of brackish waters.

to the rule; it begins to germinate while still on the tree. The radicle of the embryo pierces through the apex of the fruit, and forms a long spear-like body (fig. 57, 1). The little warty projections seen on it indicate the future side-roots. After several months the seed, or rather the young plant, parts from the parent tree, drops into the soft mud like a falling

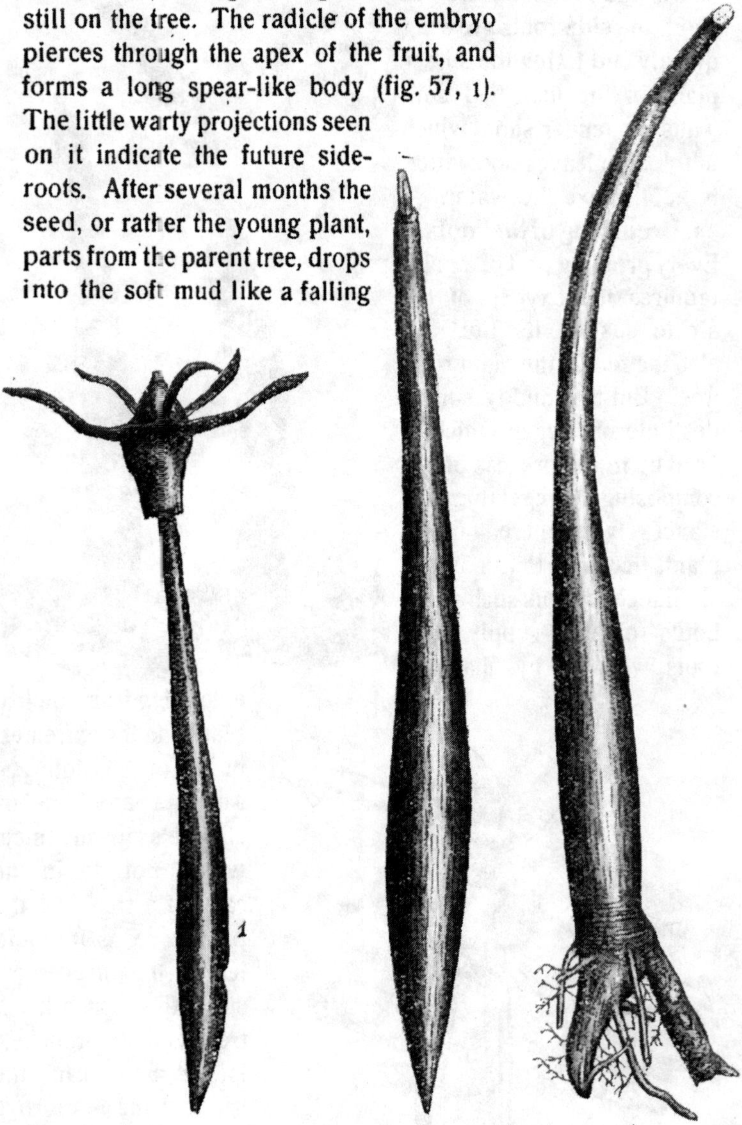

1 2 3

Fig. 57.— *Bruguiera gymnorrhiza* (Kandelia Rheedii), embryos in $^3/_4$ natural size: 1. still sticking in the fruit; 2. detached, with the plumula at the top; 3. rooted, only the basis of a shoot.

arrow and remains there upright, the side-roots growing quickly and fixing the tender plant in the unsettled soil. Thus the tender shoot which unfolds its leaves soon after, is kept above the water.

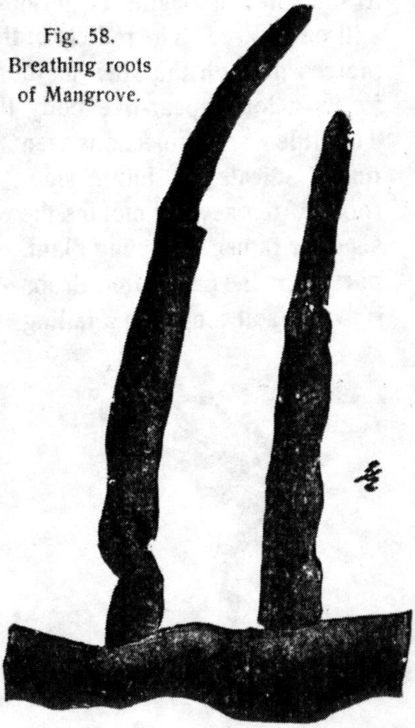

Fig. 58. Breathing roots of Mangrove.

5. Breathing of the roots.– Every growing part of a plant requires the oxygen of the air to sustain its life. So also the root of the Mangrove tree. But the muddy soil is destitute of oxygen which is used up in the process of decomposing the vegetable substances lying there. Other plants that have to live under similar conditions such as the Lotus (page 3) supply their roots with air by channels extending from the leaf blades to the extremest points of the root. But such a system of channels in the stem would not do in the case of a tree like the Mangrove, as it would lessen the mechanical strength of such a tall tree with its branches. How, then, can the roots of the Mangrove tree breathe? They simply throw up little branches above the

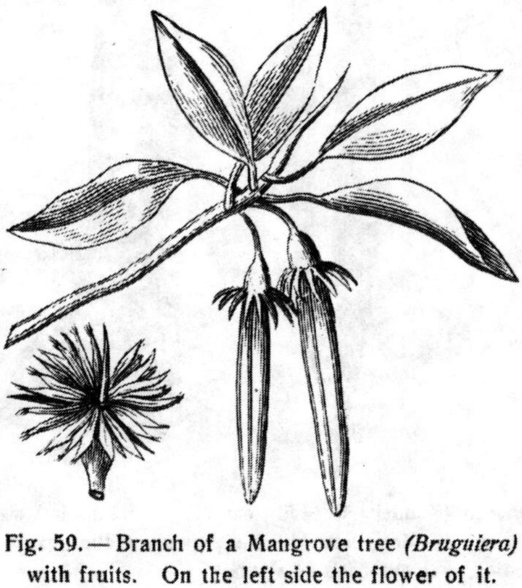

Fig. 59.— Branch of a Mangrove tree *(Bruguiera)* with fruits. On the left side the flower of it.

water level (fig. 58) containing large spaces in them and holes at their ends, by which they can get sufficient air to breathe.

6. **Flowers.**—The flowers of the Rhizaphoraceæ are radial like those of the myrtles, the calyx and the corolla being variously lobed, and the stamens numbering usually double the number of the petals (fig. 59).

13. The Cactus Family.

(Cactaceæ).

Leafless, succulent herbs. Flowers usually radial with indefinite petals and stamens. Ovary inferior. Fruit a berry.

This is an American family, but some cactuses, principally the **Prickly Pear** (*Opuntia Dillenii; Kan.* Jiḍegaḷḷi, Muḷḷugaḷḷi; *Mal., Tam.* Nāgatāḷi; *Tel.* Nāgajamuḍu) and the **Night-flowering Cactus** (*Cereus grandiflorus; Kan.* Kaḷḷi) are very widely naturalized in India. They are *desert plants* and are, accordingly, by the peculiar structure of their leaves and stems enabled to endure the longest drought. Leaves, with their large surfaces, which allow a great deal of the water to evaporate, are dispensed with and become scales which soon fall off (deciduous),—some long thorns and several short hairs standing in their axils. But the Cactus cannot do without starch to grow on, and so the thick stem always remains

Fig. 60.—Opuntia with five fruits.

green and does the work of the leaves in the preparation of starch from the air. The moisture, of which the plant gets a very scanty supply, is stored up, as in a reservoir, in the fleshy parts of the stem and evaporated very frugally, the epidermis of it being thick and almost water-tight, the stomata in it being

Fig. 61.—The Night-flowering Cactus *(Cereus grandiflorus)*.

very few, and the sap becoming a thick and slimy fluid which does not easily pass into vapour. Thus the plants are able to thrive even when everything around them is dried up. Being the only succulent things in the deserts in which they grow wild, it is good for them to be protected by sharp spines and dispense with their leaves.

There are three distinct forms of them, namely the globular (fig. 62, the central plant), the columnar (fig. 62, to the right), and the lobed or jointed (fig. 60). The Prickly Pear *(Opuntia Dillenii)* and the night-flowering *Cereus* have very showy flowers. Their fruits are soft and edible. In Mexico the Prickly Pear is cultivated for the sake of the cochineal insect that lives on it and yields a red dye. Here in India it is used for hedges.

Fig. 62.—A group of Cactaceæ.

Lately a variety of Opuntia without prickles was reared, which will, perhaps, enable us to make deserts more habitable, if the plant can now be used as fodder for cattle.

Fig. 63.—*Bryophyllum calycinum.*

One plant, often called 'Cactus' is really a Euphorbia; it has to live under similar conditions and has developed the habits and appearance of the Cactus (see *Euphorbiaceæ*).

Bryophyllum calycinum (*Kan.* Kāḍubasale) belongs to an allied family, the *Crassulaceæ*, and is also xerophilous in its habits and structure. Adventitious buds are formed in the edges of the leaves, from which young plants are produced.

SUB-CLASS 2.—GAMOPETALÆ

Plants with two floral envelopes: calyx and corolla. Corolla with united petals, the stamens generally inserted on it.

14. The Coffee Family

(Rubiaceæ).

Trees, shrubs, or herbs with opposite simple leaves and interpetiolar stipules. Flowers radial. Corolla tubular, four or five lobed; stamens, as many, inserted on the corolla (epipetalous). Ovary inferior, of two syncarpous carpels. Fruit a drupe, or a capsule, or a berry. Seeds endospermous.

The Coffee Tree *(Coffea arabica).*

(Plate No. 633.)

(*Kan.* Kāphi. *Mal.* Bunnu. *Tam., Tel.* Kāpi.)

1. Coffee is the seed of a small tree, cultivated in India, but a native of Arabia. Under cultivation the shrub is generally not allowed to grow more than six or eight feet high, but if left to itself would become a small tree.

The **Leaves** are oblong and pointed, the margins being slightly waved. They are placed opposite one another, and in such a way that every pair stands crosswise over the next lower pair *(decussate):* so also the many branches which are wiry and spreading horizontally. This ensures the advantage of the stem being equally loaded. The surface of the leaves is smooth and shining, a property which prevents too rapid an evaporation of the sap in the leaves (*cf.* Mango tree, page 25). Observe also the stipules between the petioles of the opposite leaves on either side: they coalesce into one and are termed "interpetiolar".

2. The pretty white and rose-tinted **Flowers** stand in little cymose clusters (fascicles) in the axils of the opposite leaves and have a most delicious fragrance. Calyx small. Petals gamopetalous, funnel-shaped with a short tube and four to seven or nine oblong lobes, twisted in bud. Stamens as many as petals, with lanceolate anthers attached in the middle of their back (dorsifixed) to the short filaments which are inserted at the mouth

of the floral tube between the corolla-lobes. Ovary inferior with two one-seeded cells and one bifid style.

3. **Fruit** a two-seeded drupe (commonly, but erroneously, called berry) with a blood-red fleshy exocarp crowned by the persistent calyx, and a hard parchment-like (coriaceous) endocarp enveloping each seed (Plate No. 633, 5, 6, 7, 8). The seeds are flat on their inner side with a deep inwardly curved groove (9, 10). Crush one of the seeds, sold in the bazaar as coffee-beans, and you will find that the folds consist of a horny mass and that at one end of the seed enveloped by the folds there is a minute embryo. The folded substance, then, is the endosperm of the seed.

Fig. 64.—The Coffee tree.

4. **Cultivation.**—The Coffee plant requires a well drained rich soil, such as is found in hilly forests. It grows best in a humid climate, and frost is fatal to it. In hot and dry places Coffee is successfully grown in shade. The plants are reared from seed in a nursery and, when a year or two old, planted in their permanent places in the plantation generally under partial shade. As shade-trees, such are preferred as go to enrich the soil, *e.g.*,

Bauhinia, Poinciana, Sesbania, and other *Leguminosæ.* Coffee being an exhausting crop, manuring is essential.

5. **Preparation of Coffee.**—When ripe, the fruit is gathered or shaken on cloths spread under the trees. The "berries" are then passed between rollers, which are set close enough to crush the fleshy part, but not so close as to crush the seeds. After being crushed, the pulp is washed away, and the berries, still in their skin, are set to dry in the sun. When dry they are again passed between rollers set closer than before, which now break the skin. The broken skin is blown away, and the beans are sorted and packed.

The raw beans are greenish in colour, and do not smell or taste like coffee. When coffee is wanted, the beans must be roasted, *i. e.,* placed in an iron vessel, which is turned over a fire. The roasted beans are then ground to powder, on which boiling water is poured, and then we get coffee.

This drink has a stimulating effect on the system, or, in other words, it rouses the nervous system to fresh activity, the sense of hunger is suppressed and the desire to sleep is driven away. This is due to a substance, called "Caffeine", contained in the Coffee beans. If this substance is taken in larger quantities, it acts as a poison; very strong coffee, therefore, produces pulpitation of the heart, congestion of blood in the brain, trembling of the muscles and similar affections of the nerves. Coffee is not a food in any way, but is merely a stimulant.

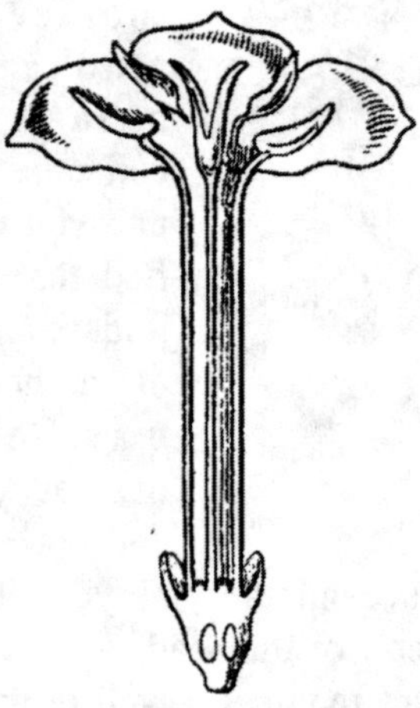

Fig. 65.—Flower of Ixora. The front part of the floral tube with the fourth lobe is removed and the ovary is cut open.

Other Plants of the Coffee Family.

The **Scarlet Ixora** (*Ixora coccinea; Kan.* Kēpala, Kisgāra; *Mal.* Četti; *Tam.* Veḍḍi; *San.* Pātalī), a very common shrub in the plains, is a general favourite because of its beautiful, scarlet flowers and edible, crimson fruit. The stem is woody and bears opposite, sessile obovate-oblong leaves, which being tough and leathery keep green and enable the plant to flower even in the hottest and driest season of the year. Stipules interpetiolar, scarious. Flowers in dense cymose clusters (corymbs) at the ends of the twigs. Calyx minute. Corolla a long slender tube that spreads into a four-parted limb. The four yellow stamens short-stalked and inserted between the lobes of the corolla. Style and ovary as in coffee. Fruit a two-seeded drupe with a fleshy scarlet exocarp.

Oldenlandia *Heyneii* (*Kan.* Urki), a small, annual herb, very common on the road-side and in open places. Stem erect, much branched, acutely four-angled. Leaves linear, entire, glabrous, one-nerved. Stipules cup-shaped, bristled. Flowers solitary or in axillary few-flowered cymes. Corolla tubular, white, limb four-parted. Stamens four, epipetalous. Ovary inferior, two-celled. Fruit a many-seeded capsule, opening across the apex.

Fig. 66.— *Mussænda frondosa.* The large white leaf is one of the five lobes of the calyx.

Mussænda *frondosa* (*Kan.* Bellotti), a straggling shrub, with acuminate leaves, and terminal compound cymes. Some of the outer flowers of the cyme with one of the five sepals enlarged, forming a large white stalked leaf (fig. 66). Corolla orange, long-tubed, limb five-lobed, star-shaped, valvate in bud. Throat of tube filled with villous hairs preventing water from wetting innner parts. Stamens five inserted in the floral tubes at different heights in different flowers: plants that have their stamens high up have short styles, and other plants with their stamens lower down have long styles on a level corresponding to the stamens of the first form: a nice arrangement for cross-pollination by insects. Fruit a two-celled berry with many seeds.

In gardens we often see the beautiful **Gardenia** with its white sweet-smelling flowers, which sometimes "double". In **Morinda** *citrifolia* (*Kan.* Maradarasina, Poppili; *Mal.* Nōṇāmaram; *Tam.* Nuṇāmaram; *Tel.* Mulugu; *San.* Dāruharidrā) the flowers stand in dense heads on a common globose receptacle. The fruits all grow together in one mass as they ripen, and look as if they resulted from a single flower (compare fruit of Ananas) forming a multiple or collective fruit.

Some of the fierce spinous shrubs in our jungles are species of **Randia** and **Canthium.** **Cinchona,** a native of Peru, yields the valuable medicine, Quinine.

15. The Composite Family.

(Compositæ).

Usually annual herbs. Leaves, as a rule, alternate, exstipulate. Inflorescence a head *(capitulum)* of many small flowers sessile on the receptacle, surrounded by an involucre of whorled or imbricated bracts, the outer florets opening first. Florets: calyx reduced to scales or bristles; corolla either tubular, or ligulate (strap-shaped); stamens five with free filaments but connate (syngenesious), introrse anthers, connective produced upwards. Ovary inferior. Fruit a dry indehiscent, one-seeded case called achenium, often crowned by a tuft of hairs, called pappus. Seed exendospermous.

The Sunflower *(Helianthus annuus).*

(*Kan.* Hottutirugana. *Tam., Tel.* Sūryakānti.)

This is an annual herb from South America which is now cultivated in gardens all over India. In some countries, chiefly in Southern Russia and in the Balkan states, it is also grown for its seeds, from which a valuable oil is made.

1. The young plants soon develop into strong and big herbs which sometimes attain a height of three yards. The thick **Stem** is branched only in its upper part. It forms a tube which is filled with loose pith.

2. The **Leaves** are large and cordate. If a thread be tied to the leaf-stalk of one of the lower leaves and then taken to the second, third, and so on, above it, one can clearly see that the leaves are *spirally arranged* around the stem (fig. 67) so that, if a small plant is looked at from above, the leaves appear in the

form of a rosette. By this arrangement each leaf gets the largest possible share of the sunlight.

The leaves *bend down to the ground* with their pointed ends, so that the *rain water* falling on them is *conducted outside.* With this arrangement the structure of the roots stands in beautiful harmony. The Sunflower plant being a tall herb, one would expect the roots to be long and strong so as to fix the plant firmly in the ground. But this is not the case. The side-roots are very short, and do not stretch beyond the ends of the leaves. As a compensation for this they are, however, very numerous and divided into so many little branches that, if the plant is taken out of the soil, the earth sticks together forming, with the roots, a compact mass which can be shaken off only with much difficulty. The water that is drained off from the centre to the circumference of the plant falls on the ends of the roots just in the same way as we have seen in the Mango tree (page 27), with one difference, namely, that, the leaves of the Sunflower plant not being close together, the rain also pours down within their circumference; consequently, we find that the tiny sucking roots are not only arranged in a ring corresponding to the outer circle of the leaf-ends, but that they are also distributed all over within that circle.

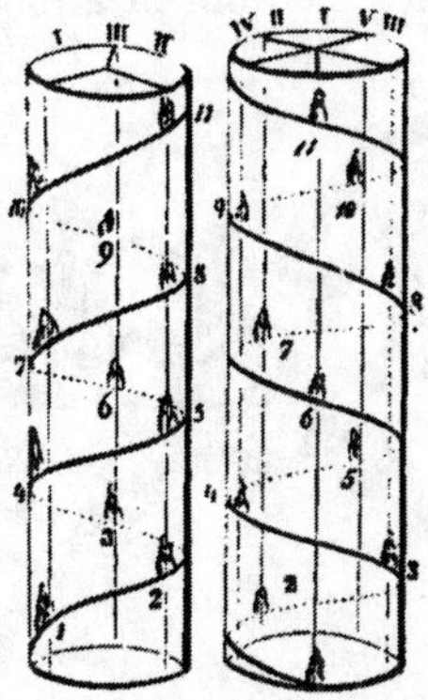

Fig. 67.—Spiral arrangement of the leaves round stems.

3. The stem and branches bear each on their tips one large **Flower,** which turns its face towards the sun (hence the name). If we cut vertically through such a "flower" (fig. 68), we can see that there are really many small flowers placed on one receptacle *(R.).* The whole is, therefore, not one flower, but an aggregate of flowers or a *head of flowers.* Hence, the order to which the Sunflower belongs is called "composite". This bunch or head of flowers is surrounded by several series of scaly leaves, imbricated one above another and called bracts *(B.),* which protect the florets under them when in bud. The florets are of two different kinds:

those in the middle have a small yellowish, tubular corolla (*tubular florets* 1, 2, 3, 4), whereas those on the margin possess a corolla stretched out in a long, yellow tongue (*ligulate florets, L.*).

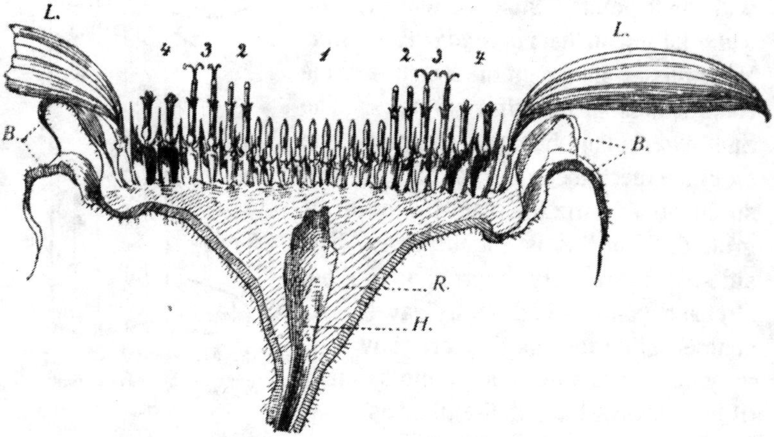

Fig. 68.—Longitudinal section of the head of the Sunflower. 1—4 Tubular florets; 1. not yet opened; 2. the pollen is pushed out of the flower tube; 3. the style protrudes and exposes its two-cleft stigma; 4. faded flowers.—*L.* Ligulate florets. *B.* Bracts. *R.* Common receptacle. *H.* Hollow part of receptacle and stem.

(a) Tubular florets.—The head of a Sunflower generally shows florets in various stages: those in the centre may be mere buds (1); then follow one or two circles of opened florets bearing clusters of pollen at their tops (2); then come florets with their forked styles visible (3); and finally, florets, which are faded, form the outer circles (4). Let us now examine one of the tubular florets (fig. 69). The ovary *(O.)*, we find, rests in a hole of the chaffy receptacle. It bears two small scaly leaves on it which represent the calyx-leaves *(C.)*. In some species of this order, as for instance the Sow-thistle *(Sonchus oleraceus)*; Lettuce *(Lactuca)*, and in *Vernonia cinerea* these calyx-leaves crown the fruit, when ripe, with a feathery ring of hairs, called pappus, by which the wind carries the seed far away. The fruit of the Sunflower have no such pappus.

The corolla of the inner florets is, as already remarked, a narrow tube. At its base there is a ball-like enlargement, and the upper part of it ends in five small and pointed teeth, indicating its composition of five connate (gamopetalous) petals. There are five stamens inserted on the corolla tube, their filaments are separated, but their anthers are joined into a tube (fig. 69, *A.*). The anthers open on their inner side (introrse) and set the pollen free. The style (*Sty.*), while seeking its way through the very narrow tube of the anthers, cannot help pushing up the pollen out of the tube where the pollen accumulates in small clusters (fig. 69, 1). Insects that visit the flower are sure to remove it with their bodies and legs. And then only the style opens its two-forked stigma in order to receive pollen from another flower (fig. 69, 2). It is a well established fact that fertilization is generally achieved not by the pollen of the same flower, but by that of another. Self-fertilization generally produces seeds of an inferior and degenerated sort. We see in this case, as well as in many other plants, a wonderful contrivance for *cross-fertilization* (compare Mussænda, page 63).

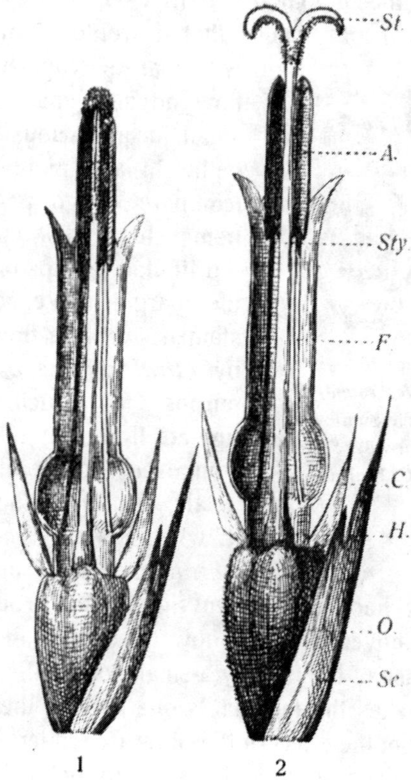

Fig. 69.—Tubular florets of Sunflower (5 times enlarged). The flower-tube is opened. *Sc.* Chaffy scale of the receptacle. *O.* Ovary. *C.* Calyx. *F.* Filament. *Sty.* Style. *A.* Tube of joined anthers (opened). *St.* Stigma. *H.* Spot where honey is secreted.

(*b*) *Ligulate florets.*—If insects have to render the flower this very important service, they must also be attracted by some means.

The flower contains honey in great quantities which is secreted at the base of the style and often fills the whole globular part of the tubular florets (fig. 69, *H.*). But a single floret would be so inconspicuous that it would hardly be noticed by the guests which are so eagerly expected. We see, now, how advantageous it is for the plant that its small inconspicuous florets should be placed together in large number forming heads of flowers (compare Mango, page 28). The effect of their being clustered is enhanced by the presence of the ligulate (strap- or tongue-shaped) florets on the margin. We seek in vain for styles and stamens in these flowers: they are sterile. But by *attracting the useful insects* they fulfil the purpose for which they were created. Each ligulate floret has five nerves showing that it is composed, like the tubular florets, of five connate petals, forming a short tube at their base.

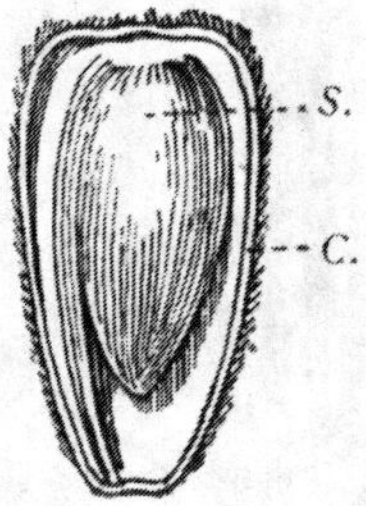

Fig. 70.—Fruit *(achenium)* of the Sunflower (opened). *S.* Seed. *C.* Fruit case.

4. The **Fruit,** when ripe, does not open, but remains shut. It is called an *achenium.* Each fruit contains one seed under its hard cover. In fig. 70 the front half of the fruit case is removed showing how the seed in the case is connected with the base of the fruit by a small cord (called funicle). When the wind shakes the tall plants one against the other, the achenia fall and from the effect of the blow are scattered around (*cf.* Poppy, page 9).

The seed contains a fatty oil in its cotyledons which serves as food for the young plant destined to grow out of it.

The Ghat Edelweiss *(Anaphalis oblonga).*

This plant may be taken as a type of the flora of high mountains. It is a small weed, branched from the root, all covered with white woolly hairs (*tomentum*), an ally of the beautiful Alpine "Edelweiss".

1. To understand its structure it is necessary to remember the *conditions of the Alpine climate.* The ruling influence of highlands is the diminished atmospheric pressure which brings in its train a number of secondary climatic effects.

The diminished atmospheric pressure is always accompanied by a decrease of the temperature in the shade, and an increase of the heat of the radiant sun. It also gives rise to great changes of temperature at day and at night. On the Kudremukh, a mountain of the Western Ghats, 6215 feet high, the thermometer stands in the month of April often at more than 40° *C.* in the sunshine and falls to 14° *C.* before sunrise. Another effect of the rarefied air is the intensity of the sunlight.

Owing to its low temperature the air cannot hold so much water-vapour as the hot air in the plains. The vapour rising with the hot air from the plains is condensed when it reaches higher altitudes, giving rise to increased rainfall on the mountains. On the other hand, the thinner atmosphere contains less vapour and is often exceedingly dry. The result is rapid changes in the humidity of the air.

Fig. 71.— Leaf-rosettes of *Anaphalis oblonga.*

The daily changes to the upward and downward currents of air cause a continual movement of the atmosphere.

The rarefication of the air, the intense radiation, the occasional dryness of the air, and the constant wind, all tend to produce intense evaporation, which every tourist will experience: everything dries up rapidly, the sweat on the skin evaporates, the skin becomes dry and brittle, and the sense of thirst is intensified.

2. *The flowers of high mountains are adapted to their special climatic conditions.* Trees and shrubs are dwarfed, having short, gnarled and bent trunks with distorted branches and scanty foliage, partly due to the increased movement of the air. Herbs are, on the whole, xerophilous, distinguished by short stems forming rosettes, large root systems, small thick leaves, often rolled up sideways and downward covering the stomata beneath, and completely covered by dense, air-entangling hairs, and the withered remains of dead foliage thickly coating the stems. Many of these low herbs are found huddled together like a flock of sheep and forming cushions like mosses, *e. g.*, the so-called **Ooty Moss** *(Anaphalis nilgherriensis).* All these are ways by which these plants are protected from loss of water; and most of them can be studied in Anaphalis.

The flowers of Anaphalis are not brightly coloured as Alpine flowers usually are, but have scarcely any colour, becoming brown and rough when withering. They are at the top of the main axis which rises from a dense rosette of leaves. Lateral branches grow from the main stem bearing also flower heads and carrying them to a higher level.

Fig. 72.—*Vernonia cinerea,* the seeds scattered by the wind.

The **Indian Cudweed** *(Gnaphalium indicum),* that grows on the hills and on the plains, is in many respects like Anaphalis.

Other Composites.

The Composite Family is the largest family of flowering plants, comprising about 12,000 known species from all parts of world. But the proportion of plants which can be used by man is comparatively small. Some are used in medicine; some are aromatic, abounding in volatile oil; a considerable number are used as salad pot-herbs.

They are divided into three tribes:—

1. Heads with the florets all similar and tubular:

The **Ash-coloured Fleabane** (*Vernonia cinerea; Kan.* Sahadēvi; *San.* Ardhaprasādana).

The **Purple Fleabane** (*Vernonia anthelmintica; Kan.* Kāḍājīrige).

The **Safflower** (*Carthamus tinctorius; Kan.* Kusubi).

Elephantopus *scaber* (*Kan.* Nelamuččaḷa), the scabrid (or rough) leaves of which appear in rosettes close to the ground at the beginning of rains, and develop flowering and branched scapes after the monsoon.

Ageratum *conyzoides* (*Kan.* Nāyituḷasi, Hēlukasa, Mūgutigiḍa), a weed frequently met with in gardens and smelling like the leaves of the Tuḷasi plant.

Sphæranthus *indicus* (*Kan.* Karaṇḍe), a prostrate weed that covers rice-fields in the cold weather.

Fig. 73.—Ligulate floret of Lettuce. *a.* Ovary. *b.* Pappus. *c.* Lower (tubular) part of corolla. *d.* Upper part of it. *e.* Joined anthers. *f.* Style. *g.* Free filaments.

Emilia *sonchifolia* (*Kan.* Ilikivi) with lyrate and hairy leaves and violet flowers.

Gnaphalium *indicum*, **Anaphalis** *oblonga*, etc.

2. Heads with the florets all similar and ligulate:

Lactuca sativa, **Lettuce** (*Kan.* Hakkarike palya, figs. 73 and 74). *Sonchus oleraceus*, **Sow-thistle** (*Kan.* Nāyi hakkarike).

Fig. 74. Fruit of Lettuce. *a.* Achenium. *b.* Pappus.

Launea *pinnatifida*, a small herb with stolons and yellow flowers, found in the sand along the seashore.

3. Heads with a ray of ligulate and a disk of tubular florets

Chrysanthemum (*Chrysanthemum indicum; Kan.* Sēvantige), a common pot-plant.

The **Sunflower** *(Helianthus annuus).*

The **Chinese Sunflower** *(Belamcanda chinensis)*, a diffuse shrub with three-lobed leaves and pretty large yellow flowers.

Zinnia *(Zinnia elegans)*, **Marigold** *(Tagetes erecta)*, **Dahlia** *(Dahlia variabilis)*, **Cosmos** *(Cosmos sulphureus and C. bipinnatus)*, all of them common garden plants.

The **Australian Daisy** *(Vittadenia australis)*, much cultivated, and also run wild at Bangalore and on the hills.

Eclipta *alba (Kan.* Garga; *Mal.* Kaiyaṇṇi), a much-prized medicinal weed.

16. The Cucumber and Gourd Family

(Cucurbitaceæ).

Climbing herbs, with hollow stems and large, rough, alternate leaves and lateral tendrils. Flowers radial, unisexual, yellow or white. Petals more or less united. Ovary inferior, three carpels parietal placentation. Fruit a berry.

The Cucumber* *(Cucumis sativus).*

(*Kan.* Muḷḷusaute. *Mal.*, *Tam.* Muḷḷuveḷḷarika. *San.* Urvārukā, karkaṭi.)

1. **Fruit and its Use.**—The Cucumber is extensively cultivated for its fruit. A cross-section through the fruit, a berry, shows three divisions, which indicate three carpels. The carpels are bent inward to the centre, where they turn again to the periphery to return once more inward towards the centre bearing on their edges the seeds and thus forming six placentas with each a group of seeds. The seeds are embedded in a sticky, jelly-like pulp.

2. **Germination.**—The seeds that fall on moist ground soon begin to germinate, the sticky flesh round them drying up and fixing the seed firmly to the ground. When germinating, the main root first appears out of the pointed end of the seed (fig. 75, 1), and sinks at once into the ground where rootlets are soon de-

* If *Cucumis sativus* is not at hand, any other of the many cultivated kinds of Gourds, Melons, or the Pumpkin, will do equally well as another type of the family.

veloped (fig. 75, 2). After that, the part of the stalk between the root and the seed-bud (hypocotyl) begins to grow, but as the root is moored in the ground and the seed-shells stick firmly to the earth, the stalk becomes a small bow, bent upwards (fig. 75, 3)

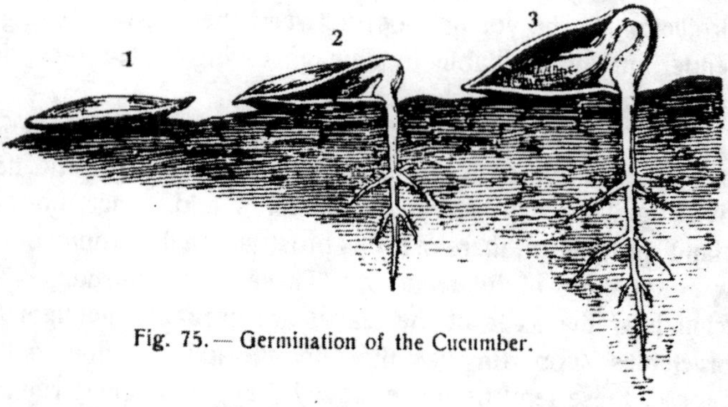

Fig. 75. — Germination of the Cucumber.

until, by its continued growth, it draws the seed-leaves out of the seed-shells. In this process the hypocotyl is aided by a swelling just above the root which keeps the lower half of the seed-case (testa) down, and disappears again after the cotyledons are liberated from the seed-case. If some seed are similarly laid on the ground, but freed from the sticky mass round them, they also germinate after a short time, but as the seed-shells are not gummed to the earth, the stalk lifts them up, and as the seed-leaves can get rid of their covers only with great difficulty, the plants may perish. This shows how important it is for the seed to be provided with that sticky pulp.

The fruit does not open by itself to let the seeds escape. For this purpose the help of man is required in the cultivated kinds, and that of animals in wild ones. Like numerous other plants whose seeds are dispersed by animals, the Cucumber has a *fleshy, edible pulp which attracts animals,* in addition to the jelly-like pulp. When, for instance, a jackal eats such a fruit, many seeds, to be sure, will be devoured together with the pulp. The number of seeds, however, being very great, this is not a serious loss. On the other hand, some seeds will stick to its mouth and legs, and will thus be spread far and wide.

The seedlings show a peculiarity, which we have already noticed in the leaves of the Bean (see page 34). They *fold up their seed-leaves* face to face *at sunset,* and expand them when daylight comes again. We have learned the importance of these movements, but here the folded seed-leaves seem to afford protection to the young shoot between them, which being very tender, might be liable to damage owing to the reduction of temperature during the night.

3. The **Stalk,** as well as the leaves, are covered with numerous *bristly hairs* as a protection against animals. The hollow, five-edged *stem is succulent and long,* and, hence, not able to stand upright. It, therefore, lies prostrate on the ground or climbs with the help of its *tendrils.* (These wiry appendages rising from near the axils of the leaves are probably metamorphosed bracteoles supporting, as they do, the axillary flowers.) The ends of these tendrils move slowly, but continuously round, like the hands of a watch. The time they require for one circuit differs and depends chiefly on the temperature. If we put a little stick in the way of the moving tendril, we can notice the following: A few hours after the tip of the tendril touches the stick, it will have formed a sling round it. Some time later we shall find the stick wound round several times, and in the course of a few days the part of the tendril between its base and the stick will be coiled up like a corkscrew. The *coiling* is always accompanied by *twisting,* and since the base and the end of the tendril are fixed before these processes take place, the directions in which the tendril coils round must

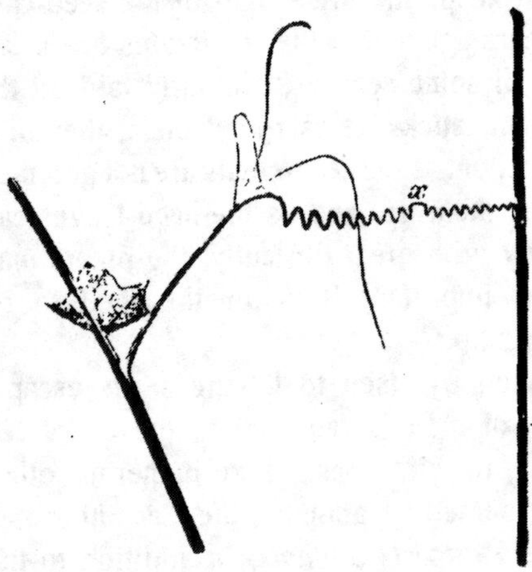

Fig. 76.—A twig of *Luffa acutangula* with tendril.

necessarily be different at the two ends. Every coil has, therefore, a turning point in the middle (fig. 76, *x*). If a tendril cannot find a support to coil round, it produces little disks from its epidermis and fastens to any flat object, or it penetrates into any crevices. In this way, the creeper fastens itself to various objects within its reach, and as the corkscrew-like tendrils act like springs, the wind or any shaking influence cannot easily tear away the plant from its support. That part of the tendril which holds the support soon hardens, thus preventing the tendril from slipping and losing its support again.

4. The **Leaves** are spirally arranged round the stem. But as a plant that lies or creeps on the ground can receive light only on *one* side of its stem, all its leaves should be directed to that side. To this end the long leaf-stalks make all sorts of turns and twists, thereby placing the leaves so that not one of them shades the other.

The leaves are *broad* and *cordate* (heart-shaped), the largest measuring about five inches each way. If we remember how succulent all the parts of the plant are and how much water it therefore requires, we can easily see the advantage the plant derives from the largeness of its leaves. Large leaves cover more ground than small ones; hence, they prevent evaporation of water from the soil in a greater measure than could be done by smaller leaves.

Large leaves are liable to be torn by the wind much more easily than small ones; and in heart-shaped leaves like those of the Cucumber the weakest part is the base. Therefore this particular part is specially strengthened: there are five or seven strong ribs issuing from the base of the leaf like fingers from the palm of the hand, and two of them form the margin of the leaves for a small stretch like the hem of a garment.

5. The **Flowers** rise singly or in small clusters from the axils of the leaves. The calyx and the yellow corolla are combined at their base. The corolla forms a five-lobed bell. So far, all flowers of the plant are alike. But if we proceed to examine their stamens and pistils, we find that most of the flowers have only stamens (fig. 77, 1), whilst some have only pistils (fig. 77, 2). The former are known as staminate or male flowers, the latter as

pistillate or female flowers. The flowers, then, are *unisexual*. But as both kinds of flowers are on the same plants, they are called *monœcious** plants. Insects carry the pollen from one

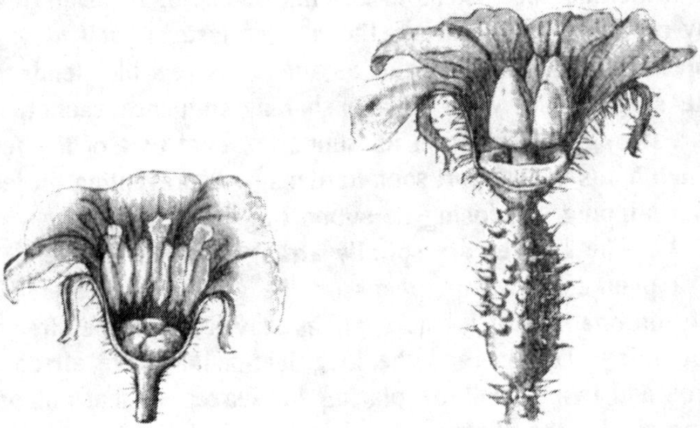

1 2

Fig. 77. — *Cucumis sativus*. 1. Staminate, 2. Pistillate flower (the front part of the corolla is removed to show the central organs).

flower to the pistil of the other. To attract them there is nectar at the base of the corolla, to reach which they must penetrate far into the flower, which is lined with thick hairs under which the nectar lies. In the act of busily seeking after the sweet liquid, the insects cannot help touching the male or the female organs, as the case may be, and so fertilize the ovules.

Other Cucumbers and Gourds.

The Gourds grow well in the warmer parts of the earth, especially within the tropics. Many of them are cultivated for their eatable fruits, and are very much alike in their general habits. They can be best distinguished by their various fruits. The commonest are the **Squash Gourd** or **Common Gourd** (*Cucurbita maxima; Kan.* Kumbaḷa; *Mal.* Čakkarakumpaḷaṅa; *Tam.* Kumbaḷam); the **Pumpkin** (*Cucurbita pepo; Kan.* Būdi-kumbaḷa; *Mal.* Kumbaḷam; *Tam.* Kalyāṇapūčuṇi); the **Water**

---

* From Greek *monos*, one, and *oikos*, a house.

Melon (*Citrullus vulgaris; Kan.* Baččaṅgāyi, Kallaṅgaḍi; *Mal.* Vattakka; *Tam.* Vattākku); the **Bottle Gourd** (*Lagenaria vulgaris; Kan.* Kahi sōre; *Mal.* Kaippačuram; *Tam.* Sūrai); the **Luffa** (*Luffa acutangula; Kan.* Hīre; *Mal.* Piččakam; *Tam.* Pīrku); the **Snake Gourd** (*Trichosanthes anguina; Kan.* Paṭla; *Mal.* Patōlam; *Tam.* Puḍōl); the **Momordica** (*Momordica charantia; Kan.* Hāgala; *Mal.* Pāval; *Tam.* Pāgal); the **Cephalandra** (*Cephalandra indica; Kan.* Toṇḍe; *Mal.* Toṇḍi; *Tam.* Kōvai); the **Common Melon** (*Cucumis melo; Kan.* Kekkarike baḷḷi, Ibbuḍlu).

Allied to the Cucumber Family is the **Passion Flower Family** *(Passifloreæ)*, likewise climbers with tendrils. Some of them, *e. g.*, *Passiflora fœtida*, grow wild, others, *e. g.*, *Passiflora palmata*, are cultivated in gardens for the sake of their peculiar flowers which have a very pretty corona of filiform appendages arising from the tube of the calyx. «The name 'Passion flower' was due to resemblances, which the mystical Fathers of the Church discovered more readily than we can. The five anthers represented the five wounds of our Saviour, the triple style the nails, the stalk of the ovary the main pillar of the cross, and the thread-like corona, the glory round His head».

The **Papaw Tree** *(Carica papaya)*, Plate No. 652, a native of America, is another plant belonging to this family. The flowers are, like those of the Gourds, unisexual. The two kinds of flowers, however, do not, as a rule, grow on the same plant, but on different plants. They are therefore called *diœcious*.* The milky sap of the tree has the peculiar property of making raw meat tender by partly digesting it.

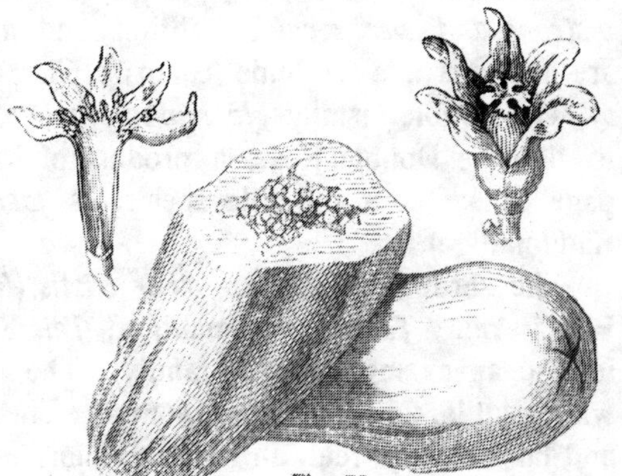

Fig. 78.
Staminate and Pistillate flowers of the Papaw tree. Below: two fruits, one opened by a cross-cut showing the seeds.

* From Greek *di*, two, and *oikos* a house.

17. The Olive and Jasmine Family

(Oleaceæ).

Trees and shrubs with opposite leaves. Flowers radial. Stamens two. Ovary superior, of two carpels.

Of this family various species of **Jasmine** (*Jasminum; Kan.* Mallige) are grown in Indian gardens on account of the sweet scent and the beauty of their flowers. They are all climbing shrubs with opposite leaves and radial flowers. The stamens and carpels are two each.

The smell of some of the Jasmine flowers is particularly strong in the evening. They also open their blossoms not in the morning, like many other flowers, but in the evening. This is, surely, not without good reason. If we look at the long and narrow tube, we may conclude that the honey at the bottom of the tube can be obtained only by insects with long tongues. Such are the moths that fly about at dusk. It is for this reason that the **Flowers** *are white,* that they *open in the evening* and *exhale such a strong and sweet scent* at that time, and that they *bend over* and are not erect like the buds (compare *Clerodendron*).

The Double Jasmine (*Jasminum sambac; Kan.* Duṇḍumallige) is, like the Double Rose, a product of horticulture (see Rose, page 45). One of the wild species is *Jasminum rigidum* (*Kan.* Kāḍumallige).

The shrub **Nyctanthes** *arbor tristis* (*Kan.* Pārijātaka; *Mal.* Pārijātakam; *Tam.* Pavaḷamalligai; *Tel.* Krishṭi; *San.* Pārijāta) is also an example of this family. The limb of the corolla is white, but its tube is orange-coloured. The flowers open at night and fall off very freely in the early morning.

The **Olive Tree** *(Olea europæa)* is not found in India, but is extensively grown in the countries round the Mediterranean Sea for the sake of the excellent oil obtained from the pulp of its fruit.

An allied family is that of the *Loganiaceæ* to which the **Strychnine tree** belongs (*Strychnos nux vomica; Kan.* Kāsarka, Hemmushṭi; *Mal.* Kāññiram; *Tam.* Eṭṭe; *San.* Kāraskarah). The tree has shining, opposite leaves and greenish flowers. The fruits are like small oranges and contain many silky seeds resembling flat, round buttons. The seeds are very poisonous, but yield a valuable medicine.

The plants of the Gentian Family *(Gentianaceæ)* are common on highlands, and known for the gaiety of their flowers. Such a beautiful flower is that of **Exacum** *bicolor*, having four white petals tipped with blue and four large yellow anthers. Exceedingly pretty also are the white, fringed flowers of the aquatic herb *Limnanthemum indicum*, very common in tanks throughout India. *Swertia chirata* (*Kan*. Kiriyātu; *Tam*. Ciriyātu; *San*. Kirātanāmā) has bitter and tonic qualities like many other Gentians.

18. The Dogbane Family

(Apocynaceæ).

Mostly shrubs, often twining, generally abounding in milky juice. Leaves opposite or whorled, entire. Flowers radial; sepals five; petals five, contorted in bud; throat often hairy; stamens five, epipetalous, filaments short; carpels two, usually free below. Fruit generally consisting of two narrow follicles. Seeds often with a tuft of hairs.

The Rose Periwinkle

(Vinca rosea).

(*Kan*. Sadāmallige, Kempu Kāsigaṇagilu.)

1. This is a small shrub found everywhere in Indian gardens. It flowers throughout the year. Drought stops the growth of plants, but it does not seem to affect *Vinca rosea*, which remains green and flourishing when everything else is withering for want of water. What makes this little plant so hardy?

If you pull the plant up, you will notice that its **Roots** are long

Fig. 79. — The Rose Periwinkle *(Vinca rosea)*.

and extend beyond its branches. This enables it to get water from a greater space than most plants of its size can.

The **Stem** of Vinca does not grow high, but has numerous long branches, the tips of which only are erect. Small plants, like Vinca, generally have a herbaceous stem which does not last for more than one season. Vinca has a strong woody stem, in which it can store up food stuff and moisture for the time of need. This stem, moreover, is covered with *a very tough and leathery bark* which will not easily allow the moisture contained in its inner layers to evaporate. The sap in the stem is, besides, slimy, and slimy fluids, as a matter of fact, dry up very slowly (compare *Cactus*, page 58).

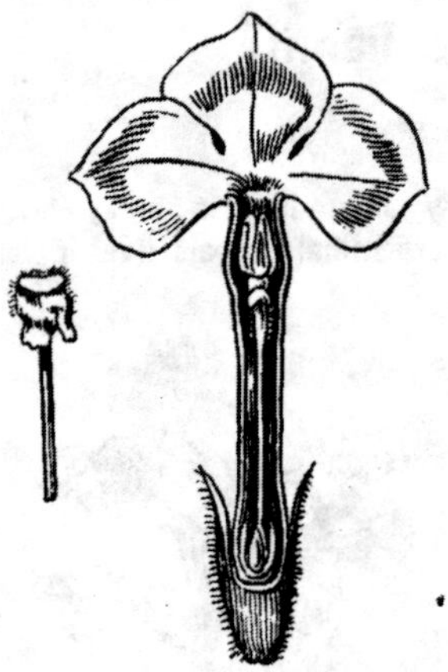

Fig. 80. — Flower of Vinca. The two front lobes and half of the floral tube are removed to show the stamens and the style. On the left side: the upper part of the style with the hourglass-like stigma.

The elliptic **Leaves** of Vinca are placed opposite. They are shiny above and provided with a thick epidermis, which conditions reduce the evaporation (see Mango, p. 25). Touch the leaves and you will see that, though they shine and appear glabrous or smooth, there is a fine *coat of down* all round the epidermis. This also helps to reduce the evaporation.

2. The **Flowers** grow in pairs in the axils of the leaves. The calyx is divided into five filiform segments. The corolla consists of a long cylindrical tube spreading at its upper end into five broad limbs, which are contorted in bud, but are at right angles to the tube (salver-shaped) when open. The mouth of the tube, tinged with dark crimson, is slightly raised, surrounded by a corona of hairs, and very narrow. But a little below, the tube widens, making room for five sessile stamens, which form a cone under which the hourglass-like stigma of the long and slender style is situated. The latter rises from the combined tip of two apocarpous seed-vessels at the bottom of the floral tube (fig. 80).

3. **Pollination.**—Between the two seed-vessels there is a gland on each side which secrete nectar. Only insects with long tongues can get the nectar at the bottom of such long flower-tubes. Such are certain night moths. They, in their turn, have to pollinate the flowers. The anthers are inclined over the flat top of the style and deposit their pollen there. But this part of the style is not receptive. The only part in which the style can receive pollen is the collar or seam of the hourglass-like stigma, which is sticky. Now, there are five little openings between the five anthers through which the insect must stretch its tongue into the flower-tube. While doing so, the tongue brushes against the sticky sides of the stigma and deposits there any pollen that it may have brought from another flower. When the insect draws its tongue back out of the tube, it touches the pollen-masses lying loosely on the top of the stigma which readily adhere to the sticky tongue. Thus the pollen of one flower is carried to the style of another, and self-pollination is avoided.

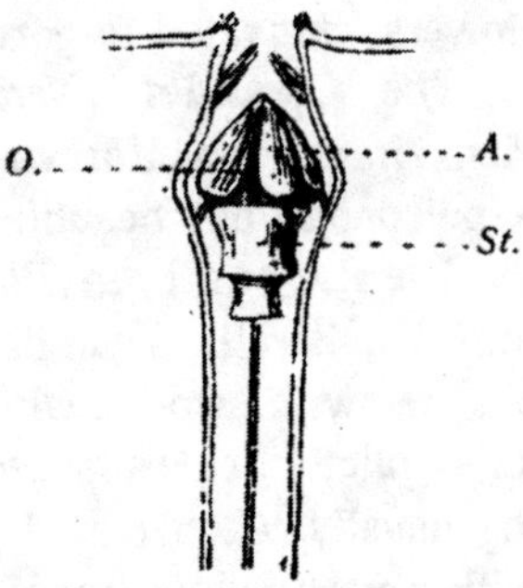

Fig. 81.—Longitudinal section of floral tube of Vinca. *A.* Anthers. *O.* Opening between anthers. *St.* Stigma.

4. The **Fruit** consists of two erect and cylindrical follicles, dehiscing each in its ventral suture. Seeds numerous and small.

Other Apocynaceæ.

The **Corrinda** (*Carissa spinarum; Kan.* Garji, Korinḍa; *Mal.* Karanṭa; *Tam.* Kali; *Tel.* Kalivi) is a common shrub, with milky juice, spreading branches and opposite, axillary, often forked spines (representing metamorphosed branches). Stipules minute, interpetiolar leaves subsessile, elliptic, entire, glabrous, coriaceous. Flowers white, sometimes with a tinge of pink, fragrant, in cymes. Parts of flowers similar to those in Vinca. Fruit a two-seeded berry with a very sticky pulp, edible.

Alstonia *scholaris* (*Kan.* Hāle; *Mal.* Ērilapāla; *Tam.* Ēḷilappālai; *Tel.* Ēdakūla; *San.* Jīvani), a large tree with a bitter bark valued in medicine. Base of trunk often buttressed by plank roots; branches spreading horizontally in tiers of whorls. Leaves in whorls of five to ten.

Tabernæmontana *coronaria* (*Kan.* Nandibaṭlu; *Mal.* Tagaram; *Tam.* Nandyāvaṭṭam; *Tel.* Nandivardhanamu; *San.* Vishṇupriya), pure, white flowers, often double, fragrant at night.

Allamanda *cathartica* (*Kan.* Arasina-hū, Seitāna-hū, Kēḷa), a scrambling shrub with large, grotesque, bell-shaped, yellow flowers, cultivated in gardens and run wild in many places.

The **Oleander** (*Nerium odorum; Kan.* Kaṇagilu; *Mal.* Kaṇāvīram; *Tam.* Karavīram; *Tel.* Kastūripaṭṭe; *San.* Karavīrah), a poisonous but beautiful shrub. Leaves in whorls of three.

The **Pagoda Tree** (*Plumieria acutifolia; Kan.* Kāḍusampige; *Mal.* Veḷuttaraḷi; *Tam.* Īḷattalari; *Tel.* Aḍavigannēru), a deciduous tree with swollen white trunk and branches, full of a sticky milky juice. Leaves lanceolate, acute, in terminal tufts. Flowering usually when out of leaves. Flowers fragrant and white, with a pale yellow throat. A native of Mexico, but cultivated in India from time immemorial; does not produce seed.

Cerbera *Odollam* (*Kan.* Čaṇḍe; *Mal.* Utaḷam; *Tam.* Kāṭaraḷi) grows in salt swamps, adorning them with its thick foliage and its large bunches of white flowers. The fruit, a drupe, is, like the cocoanut, beautifully adapted for dispersion by running water. When a fruit drops into the water, the outer pulp decays, the fibrous covering serves as a swimming belt and the inner hard covering as a protective coat for the seed.

19. The Milkweed Family

(Asclepiadaceæ).

Shrubs often twining, usually containing a milky juice. Leaves entire, opposite. Flowers radial; petals five, contorted in bud; stamens five, inserted on the base of the corolla. Anthers coherent enclosing the stigma, filaments in most genera connate into a staminal tube, the end of the style being a broad disk with the stigmatic surface underneath. Ovary of two carpels. Fruit two follicles. Seeds with a brush of hairs.

The Madar *(Calotropis gigantea)*.

(*Kan.* Ekka. *Mal.* Erikku. *Tam.* Arkkam. *Tel.* Arkamu. *Hind.* Madār. *San.* Arkāh.)

1. The **Stem** and the **Leaves** abound in *milky juice,* which protects the plant in many respects. It is acrid and poisonous and makes the plant distasteful to cattle. The viscid resin contained in it causes it to clot readily so that wounds, through

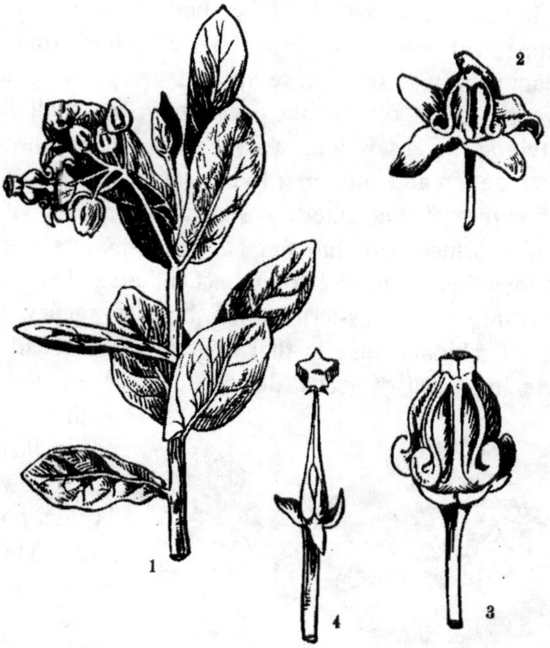

Fig. 82.—The Madar *(Calotropis gigantea)*, reduced.
2. A complete flower. 3. The flower stripped of the corolla showing the curved appendages round the seed-vessels. 4. The two seed-vessels, stripped of the appendages.

which it oozes out, are soon shut up, thus preventing germs of decay from entering into the body of the plant. Its bark contains a very strong and useful fibre. The soft, white, woolly tomentum on the stem and on the underside of the leaves affords the plant protection against the withering influence of dry winds. Owing to this and the thickness of the epidermis, the plant is, in spite of the size of its leaves, in a position to remain green even during the hottest and driest part of the year. Leaves pretty large, opposite, oblong-obovate with auricled base and short stalks.

2. The **Flowers** grow in large cymose umbels. Sepals five. Corolla bluish, five-lobed, bell-shaped; lobes valvate in bud. Of peculiar interest is the structure of the essential or reproductive organs of this flower. What strikes us first when we examine it, are five large, wax-like, bluish bodies, curiously recurved, alternating with the corolla lobes and radiating from the staminal column (fig. 82, 3). These appendages form what is called the corona. If we remove them carefully, we shall find in the cavity two seed-vessels with a style each. These, however, are united at the top and support a five-rayed, flat stigma (fig. 82, 4). At each corner of this cake-like stigma a wiry, brown body can be seen, to which two club-shaped pollen-masses are attached, each of them belonging to two distinct anthers. This is evidently a contrivance for cross-fertilization by the agency of insects (compare Orchids). Insects that alight on the broad stigma of the style in search of nectar detach some of the pollen-masses with their feet, and carry them to another flower, which is thus pollinated.

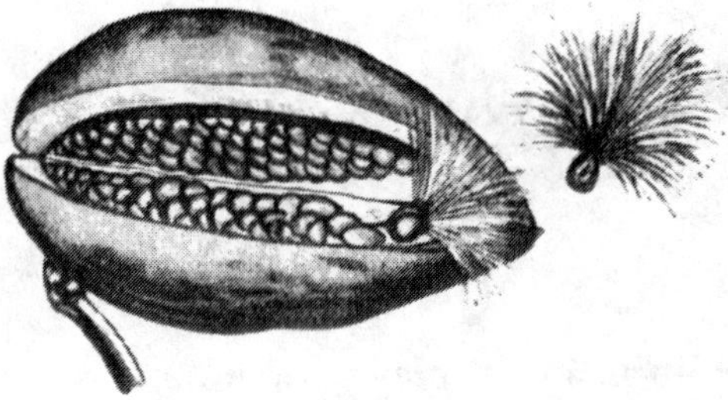

Fig. 83. — Follicle of *Calotropis gigantea.*

3. The **Fruit** is made up of two large, ovoid, dry follicles, each splitting up in one line (the inner or ventral suture) and containing numerous flat seeds packed together in beautiful order. One of the two follicles is sometimes absent by abortion. Each seed is crowned with silky hair. When ripe, the feather expands like an umbrella, and caught by the wind, may be

carried away to a great distance with the seed attached to it beneath. Compare this pappus with that found in some Composites (page 70). In this case the silk floss is formed inside the carpel, whereas in the Composites it is a growth on the top of the carpels.

4. Most of the **other Asclepiadaceæ** are climbers. One of them, **Hemidesmus** *indicus* (*Kan.* Nāmadabēru; *Mal.* Nannāri; *Tam.* Nannāri; *San.* Bhadravalli), affords the so-called Indian Sarsaparilla. **Dæmia** *extensa* (*Kan.* Juṭṭuve, Pettatajaṅk; *Tam.* Belaparti) is very common and possesses cordate leaves and greenish flowers.

20. The Bindweed Family

(Convolvulaceæ).

Herbs, rarely shrubs mostly twining, with alternate, exstipulate and usually cordate leaves. Mostly twiners. Many with milky juice. Inflorescence cymose. Flowers radial. Corolla funnel-shaped, folded inwards (in duplicate) and twisted (convolute) in bud. Stamens five, epipetalous, often of unequal lengths. Ovary superior, of two carpels, usually with two large seeds in each cell. Seed endospermous. Cotyledons folded.

The Elephant Climber *(Argyreia speciosa).*

(*Kan.* Samudrapāla. *Mal.* Samudrajōgam.)

This is a huge climber, commonly found in thickets and jungles, especially near the sea. Stem and lower side of leaves covered with white, silky hairs. Leaves large and cordate. We choose this plant as a type of the lianas, a name given to climbing and winding plants in tropical forests.

1. **Why the Elephant Climber climbs.**—Plants cannot live without light. If we grow any plant in a flower-pot and keep it in a room, it will invariably stretch its branches towards the window, eagerly seeking the light. If it cannot get sufficient light, it will etiolate, *i. e.*, grow white from absence of the normal amount of chlorophyll in its leaves, and the stalks become unnaturally long, producing small and bent leaves, but no

flowers. The plant is sickly for want of the necessary light. A dense forest produces the same effect. There is so much shade that hardly any herb can grow in it. How many plants die there for want of light! How many seeds fall there to the ground and germinate, but cannot grow further, perishing miserably in the dark thicket!

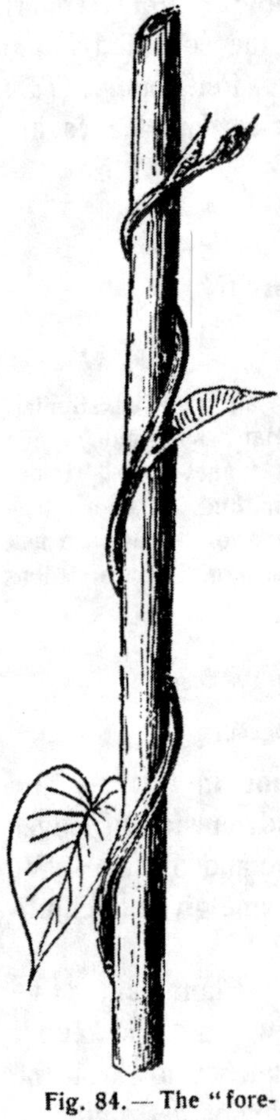

Fig. 84.— The "fore-running tip" of *Argyreia.*

The seed of a liana has to germinate under such disadvantages and would certainly suffer the same miserable death, if it were not provided with certain qualities that other plants lack. Above all, the liana is distinguished by an unusually quick growth, by which it brings its foliage to the top of trees in a very short time. The darkness and the damp air in the jungle are advantageous to quick growth. It is as if the plant in its youth had no other aim than this, and as if everything else about its growth was subservient to this one end: it does not form many leaves; it does not produce branches; the stem does not grow in width; it only grows in length. The stem thus becomes so slender and weak that it cannot stand upright by itself. What may now be its fate? Shall it yet die in the struggle for existence?

It would certainly die, if it had not the wonderful ability to use as its support the very giant tree in its vicinity which threatened to be the cause of its death and to choke it. The Elephant Climber climbs the tree by winding round it. The leafless tip of the slender stem moves spirally until it finds something to grasp, round which it then winds in the direction opposite to that in which the hands of a clock move. If the stem cannot find any support, it sinks down to the ground, striking new roots at its nodes and rising again. If no vertical support can be obtained, the plant does not thrive well. Supports standing up vertically are preferred to any others, for they enable the plant to come to the light in the shortest time.

2. **The "Forerunning Tip" of the Lianas.**—The tip of the climber is of particular interest being characteristic of all climbers. There are only a few very small leaves or rather leaf-buds on it at large intervals, and these buds do not unfold for a long time, *i.e.*, till the part of the stem above them has taken a firm hold. The advantage a climber derives from this constitution of its tip is evident. It keeps the stem light and movable, and enables it to slide through any hole in the thicket through which it seeks its way up to the light. It has rightly been called the "forerunning tip" of the lianas.

Not all the lianas are *Twining Climbers*. Some climb by sliding with their tips through the branches of other trees and growing spreading branches which lean against their supports without actually clinging to them. These climbers are often furnished with spines and thorns which they use to prevent their gliding back. Such scramblers are the Rose (see page 45), the Rangoon Creeper *(Quisqualis indica)*, the Bamboo, etc. Other climbers produce rootlets along their slender stems by which they cling to their supports as by a thousand little fingers. Such *Root-Climbers* are the Pepper vine and many genera of the family of the Aroideæ, *e. g.*, *Pothos scandens* (*Kan.* Aḍakebiḷu; *Mal.* Ānapparuva). Most climbers, however, develop special organs for climbing, namely tendrils, by means of which they seize their supports. *Tendril Climbers* are the Pea, the Cucumber, the Grape vine, etc.

3. **The Thirst for Light in Plants.**—In the second part of this book (Assimilation) it will be shown why plants require the light of the sun for their growth. The amount of light required is, however, not the same for all plants nor is it the same for the different parts of a plant. Plants generally protect their young buds and shoots against the direct light of the sun, as for instance the Elephant Climber, by keeping them folded for some time and by a dense coat of hair on the outer side, or as the Mango tree, by providing a red pigment in the outer cells and by letting the leaves hang vertically. This is because intense light destroys the green grains in the cells of the leaves. The glaring light

of the midday sun is too strong not only for young shoots, but also for grown-up plants, and many of them, therefore, protect themselves against it by folding their leaves at that time (Acacia), others by reflecting it from the shining surface of their leaves (Mango), others again by seeking shelter in the shade of other plants. Thus we find the plants growing in a wood to be arranged in about five different tiers according to their several need of light. The full sunshine is enjoyed by the trees and the lianas at their tops. Less light is wanted by shrubs which keep below the trees and by epiphytic plants which perch on the branches and trunks of the trees. In their shade grow herbs and ferns. Below these crouch the mosses and other cryptogams. The lowest and darkest place is occupied by the fungi, which require hardly any light, leading a parasitical life in the leaf-mould or humus at the bottom of the forest.

4. **Flowers.**—On reaching the top of trees the Elephant Climber assumes another mode of growth. The tip need no longer grow so quickly, for it has brought the plant up to the light. Branches are developed and flowers make their appearance in the axils of the leaves. They are arranged in three-forked cymes, supported by large, white, deciduous bracts. The calyx consists of five imbricated sepals of unequal size, and the corolla of five gamopetalous petals, forming a large funnel or bell light-pink at the border and dark-violet within. The lobes are folded inwards and twisted in bud, and the marks of the five folds are visible in the different colouring of the corolla. A little above the base of the flower-tube rise five stamens of unequal lengths leaving five narrow holes at their base, the entrances to a little chamber in which we find the ovary tapering into a long style that rises with its two-lobed globose stigma to the level of the anthers. Pull the corolla out of the calyx, and you will see the lower chamber at the base of the corolla. The ovary is surrounded

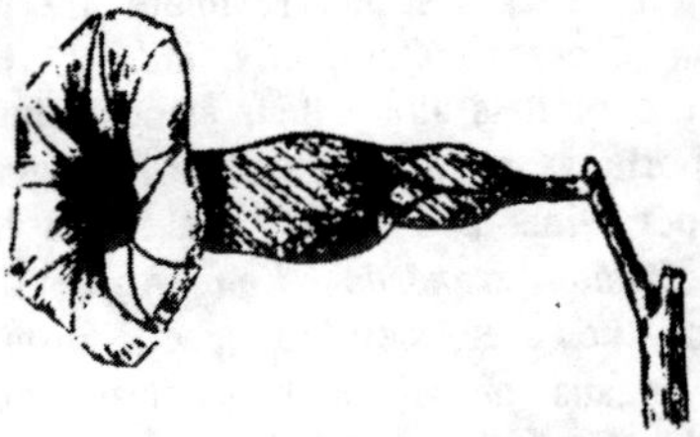

Fig. 85.—Flower of *Argyreia.*

and covered by a pad-like body secreting nectar. Butterflies which wish to sip from the nectar, must needs thrust their tongues through one of the five narrow passages formed by the bases of the stamens, and thus come in contact with the pollen to pollinate another flower when they go to visit it. Towards evening (find the exact hour) the petals fold up like the folds of a fan, their visitors having also retired.

The fruit is a berry with four large seeds.

Other Bindweeds.

Ipomœa is a very characteristic genus of this family. It has a dry dehiscing fruit, which distinguishes this genus from the genus *Argyreia.*

The **Moon Creeper** *(I. bona-nox)* has long-tubed white flowers, opening at night.

The **Needle Creeper** *(I. quamoclit),* a very common garden weed, has pinnatisect leaves, the segments being needle-shaped.

Fig. 86.— *Ipomœa batatas.*

The **Sweet Potato Plant** (*I. batatas; Kan.* Sīgeṇasu) is cultivated for its tuberous roots. *I. eriocarpa* (*Kan.* Muḷḷibaḷḷi;

Tu. Kuḍuta Tammale) is often found in Horse-Gram fields, and *I. sepiaria* in hedges, the latter having dark blotches about the midrib of its cordate and acuminate leaves. The **Goat's-Foot Creeper** (*I. bilobata; Kan.* Aḍumbu) is a xerophilous creeper on sandy seashores. It is remarkable for its two-lobed leaves joined on the inner edge like *Bauhinia* (page 41).

Evolvulus *alsinoides* (*Kan.* Vishṇukrānti), one of our loveliest flowers in the grass, has a prostrate stem with beautiful, sky-blue, short-tubed flowers. It is a xerophilous plant covered all over by long, air-entangling hairs.

21. The Nightshade Family

(Solanaceæ).

Herbs or shrubs, many of which are poisonous. Leaves alternate, exstipulate. Flowers radial. Corolla funnel or bell shaped, lobes five, folded inwards. Ovary superior, of two carpels. Fruit a berry or a capsule. Seeds numerous, endospermous.

(a) The Potato (*Solanum tuberosum*).

(Plate No. 622.)

(*Kan.* Uraḷagaḍḍe. *Mal.* Uraḷakiḷaṅṉu. *Tam.* Uraḷakiḷaṅgu. *Hin.* Baṭāṭā.)

1. **Importance of the Tuber.**—The part of the Potato plant most often seen by us is the *tuber.* This grows in the ground, but is not the root of the plant. For, if we closely examine the tubers, we shall find buds, or as they are commonly called "eyes", and also scale-like leaves which never occur on roots, but only on parts of the stem, and we thus see that the *tuber is only a swollen stem growing underground.* In full-grown potatoes the scales are not present, but there are small curved lines under the eyes. These lines are the scars of the scales. If the tubers are planted, new plants may grow out of each of these buds. Even if one of the latter is cut out and planted in the soil, it may grow into a new plant. From this it is evident that the tuber is of great importance to the life of the Potato plant.

Let us examine the tubers a little more closely. If we take two potatoes of equal size, peel one of them and expose them both to the sun, we shall after some time notice that the one that was peeled begins to shrivel, whereas the other remains unchanged. The former has lost a great deal of the water it contained. If you plant it—of course, without removing the buds,—no plant will grow out of it, for the buds are withered. So we see that it is the *skin of the tuber* which *protects it from withering.* This skin, we are told, consists of the same substance as cork. And we know that cork is almost the best material available for preventing the evaporation of liquids which are kept in bottles. Besides, the coat of cork serves as a protection against any hurtful influence from outside. Its bitter taste, for instance, saves the tuber from attacks of insects.

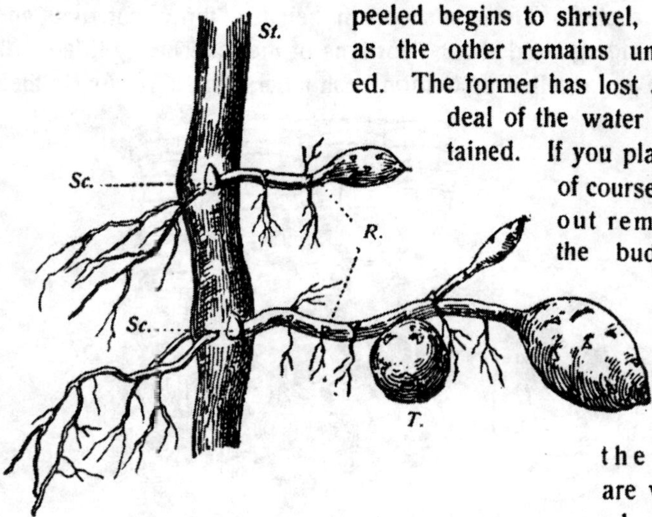

Fig. 87. — Formation of Potato-tubers. *St.* Stalk. *Sc.* Scaly leaves. *R.* Runners. *T.* Tubers.

It is by means of these tubers that *the Potato plant can endure the hot and dry season.* At the end of the season the plant is full grown and has formed flowers, fruit, and seeds, as well as a number of tubers under the ground attached to the mother plant by string-like, horizontal runners or stolons (fig. 87, R.). It will now, for want of moisture in the soil, wither and die down to the tubers which, protected under the ground, preserve the germ of life in their buds. In the following year, when the soil gets moist again, the buds begin to grow just as they do on any branch. The shoot has no roots at first and must, therefore, get all its food from the supplies stored up in the tuber, and this causes the tuber to shrivel up and die, as the food stored in it is exhausted. The new shoots in their turn throw out roots and other underground shoots, portions of the latter being filled with starch and swelled up to form fresh tubers (fig. 87). As, in their

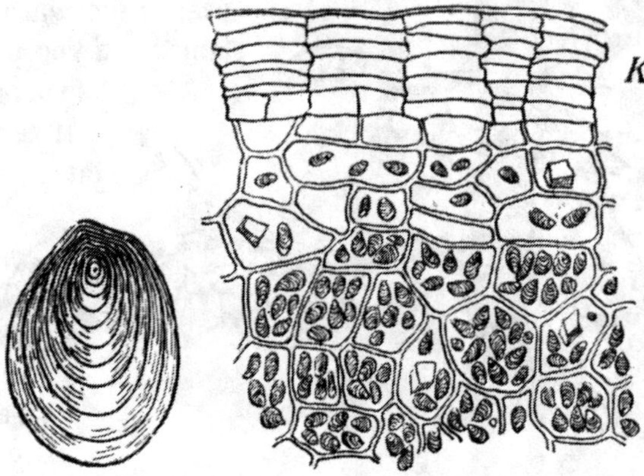

Fig. 88. — Section through the outer portion of a Potato. *K.* Cork-cells. *St.* Cells containing starch-grains (140 times enlarged). To the left a starch-grain showing its stratified structure (500 times enlarged).

uncultivated state, the tubers of the Potato plant remain in the ground and give rise to a large number of new plants, it is of great advantage to the new generation that the tubers are produced at the ends of runners, and are thus removed to a distance from the mother plant.

We see now clearly that the *potato-tuber is a store of food for the new plant.* This food consists mainly of starch, which is also one of the principal food-substances of man. As the potato-tuber contains no noxious properties and is easily obtained in large quantities, it has become one of the chief vegetables we eat. It is now cultivated nearly all over the world, but does not grow well in the tropics.

2. **Leaves, Flowers and Fruit.**—The stem of the Potato plant does not, as a rule, grow higher than about one foot and a half. The *leaves* are large and interruptedly pinnate, the leaflets being

not all of the same size and there being smaller pinnæ between the larger ones. The leaves contain a poison which is also present in all other green parts of the plant and especially in the berry.

The **Flower,** like the flower of the Chillies or of the Brinjal, consists of a calyx with five segments, a disk-shaped corolla with five lobes, alternating with the sepals, and five yellow stamens, again alternating with the lobes of the corolla (fig. 89). The large pollen-bags are united at their ends so as to form a cone surrounding the pistil (Plate No. 622, 5 and 7). The anthers open by pores at their upper ends.

Fig. 89. — Floral diagram of Solanaceæ (*Petunia*).

The **Fruit** is a round, green berry with many seeds in two cells (Plate No. 622, 6 and 9) which, however, cannot be depended on to produce plants that will give fine tubers. They are, therefore, not used except by nursery men who hope to obtain new varieties.

(b) The Tobacco Plant *(Nicotiana tabacum).*

(Plate No. 621.)

(*Kan.* Hogesoppu. *Mal.* Pukayila. *Tam.* Pugaiyilai. *Hin.* Tambāku.)

1. This plant has come from America, like the Potato, and is now cultivated in many districts of India.

All parts of it are covered with sticky, glandular hairs which keep off animals. The very large **Leaves** decrease in size towards the upper part, thus giving the lower leaves the share of sunlight which they need for the proper exercise of their functions. You will notice, too, that the leaves are almost all bent down at their tips. The plant has a deep, vertically growing taproot with side-roots growing horizontally. The latter, however, do not go beyond the circumference of the leaves and they, therefore, have their tender sucking parts just below the tips of the leaves. We see now clearly why the leaves are bent down. When it rains, all the water does not run along the leaf-stalks to the inner part of the plant, but a great deal flows outwards to where the tips of the roots are waiting for their nourishment (see Mango tree, page 27). The rest goes into the soil, where it is protected from evaporation and can sink in to the deeper lying roots.

2. The stem bears at its end great panicles of tubular **Flowers,** which are either white or red. The five stamens are inserted in the tube of the corolla between its lobes (fig. 90, 2). The **Fruit** is a capsule formed of two carpels (fig. 90, 5). When ripe, it opens in two valves (fig. 90, 4) to let the small seeds escape.

The cross-section of a seed (fig. 90, 7) shows that the embryo is embedded in a separate tissue, called endosperm, which forms a nourishing substance for the plantlet when it begins to grow. The embryo is curved.

3. In growing the plant for **Tobacco** the stem is nipped off when it reaches a height of about twelve inches and is not allowed to flower. Why? Evidently to aid the formation of the leaves out of the material that might be wasted in the production of flowers and seeds. When the leaves are fully grown, the plant is cut down and left in the field for several days, after which, early in the morning when there is still dew on them, they are removed in bundles of forty or fifty leaves. If the leaves are too dry and there is no dew on them, water is sprinkled on the leaves before removal. These bundles are then put in a stack where they are frequently rearranged from top and bottom to middle and from middle to outside. During this process of curing a fermentation by the agency of certain bacteria takes place. After about two months the leaves are ready for smoking, chewing or making into snuff.

Tobacco contains a poison, called *Nicotine,* of which a single drop suffices to kill a dog. Continuous excessive use of tobacco produces heart and bowel diseases, and can bring on the entire ruin of the body. For children tobacco, taken even in small quantities, is a dangerous poison.

(c) Other Nightshades.

The **Brinjal** (*Solanum melongena; Kan.* Badane; *Mal.* Vaḻutina; *Tam.* Vaḻudalai; *Tel.* Vaṅkāyi; *Hin.* Baiṅgan) produces the well-

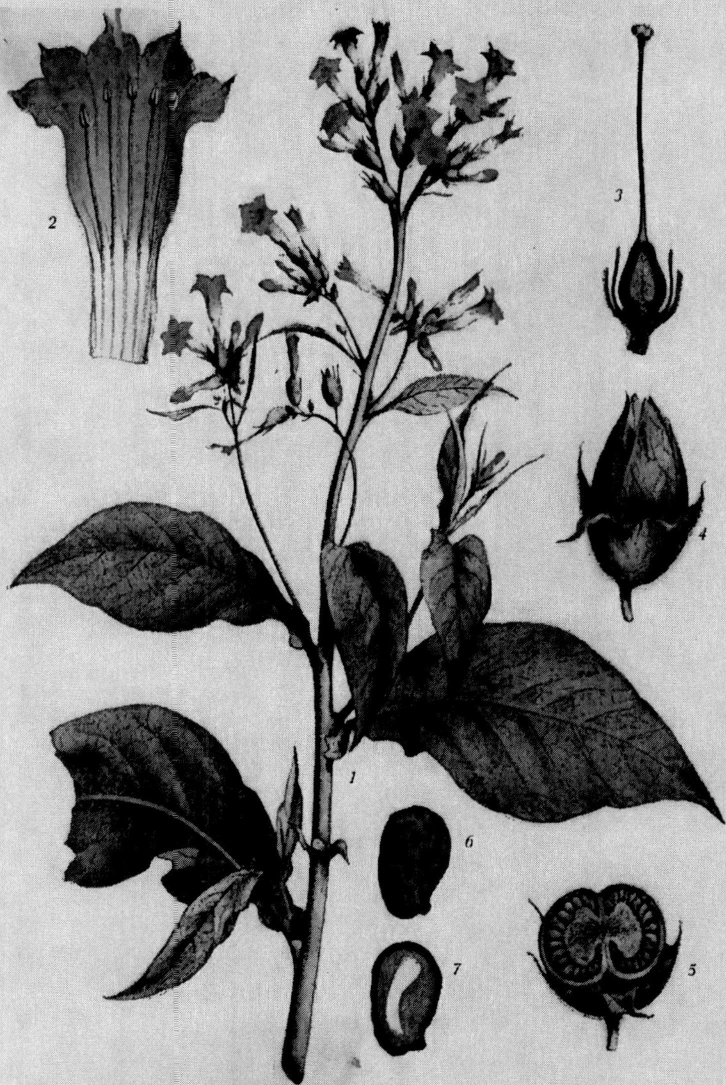

Fig. 90.—TOBACCO (*Nicotiana tabacum*).

2. Flower laid open. 3. Pistil. 4. Ripe capsule. 5. Transverse section of capsule. 6. Seed. 7. Section of it, showing cotyledons and endosperm.

Fig. 91.—CHILLI PLANT (*Capsicum frutescens*).

known egg-like fruit. The **Tomato** (*Lycopersicum esculentum*—Plate No. 635—*Kan.* Čapparabadane; *Mal.* Pētakkāḷi; *Tam.* Erumaittakkāḷi; *Tel.* Takkāḷi) is also cultivated. The so-called **Cape Gooseberry** (*Physalis peruviana; Kan.* Boṇḍuḷa; *Mal.* Moṭṭāmpuḷi) is very common in India. Its calyx is inflated, and wholly covers and protects the edible orange-coloured fruit.

Fig. 92.— The Thorn-apple *(Datura stramonium).*

The **Chillies** (*Capsicum frutescens*—Plate No. 635—*Kan.* Meṇasu; *Mal.* Pareṅgimuḷakụ; *Tam.* Miḷagāyi; *Tel.* Mirapakāyi; *Hin.* Lālmirči). The scarlet fruits of the Chillies are used as a condiment, and the plant is, therefore, widely cultivated. It is interesting to study the formation of its branches. The plant early develops its first terminal flower. Below the flower the stem branches into two forks, each of which produces again a terminal flower, the stem again forking into two branches below it, and so on. The flowers are similar to those of the Potato plant (see page 93), and the fruit is a long, dry berry formed of two carpellary leaves.

Several species of Solanum are growing wild, and among them, the commonest, perhaps, is *S. xanthocarpum,* an annual weed, prickly all over and producing yellow berries, as its name says.

The **Thorn-apple** (*Datura stramonium; Kan.* Dattūra, Ummatta; *Mal.* Ummattam; *Tam.* Ummattai; *Tel.* Ummetta) is a common weed with a strong, disagreeable smell. Its large leaves are deeply toothed, the two sides generally being unsymmetrical. They are spirally arranged round the stem and its numerous branches, the formation of which is the same as in the Chilli plant. The corolla of the flower forms a large funnel of a pure, white colour. It opens at nightfall and exhales a strong smell which, like the white colour, attracts moths which transfer the pollen from one flower to another. The capsules open in four valves, the ovary being four-celled by the inflection of the midrib of each of the two carpels so as to meet the placenta.

22. The Butterwort Family

(Lentibulariaceæ).

Carnivorous, aquatic and marsh plants. Flowers zygomorphic. Corolla two-lipped and spurred. Stamens two. Ovary superior.

The Bladderwort *(Utricularia stellaris).*

The Bladderwort is a little plant with yellow flowers, floating in stagnant water and very common in our tanks and wells. It does not root in the mud nor stretch its leaves above the level of the water like the Lotus and, therefore, has to find some other means of existence.

1. **Absorption of Water and Mineral Food.**—Terrestrial plants suck up water and mineral food by means of their roots. Aquatic plants that are wholly submerged in water absorb water

on their whole surface, the epidermis being very thin. Different salts being generally dissolved in water, the plants obtain their mineral food from the water they absorb. This being so, they require no special organs of absorption, and roots are dispensed with. Therefore the Bladderwort cannot exist in running water, but only in pools and ponds with stagnant water.

2. **Absorption of Carbonic Acid Gas.**—The Bladderwort is green; which shows that it requires carbon dioxide to form starch for its growth. This gas is found in water also, though in very small quantities. To get a sufficient supply of it, the plant has to increase its surface enormously by dividing the leaves into hundreds of parts. Take a potato and cut it into many thin slices: you will see how the surface of the piece is thus greatly increased. The greater the surface of the leaves, the more the area available for absorbing food. The leaves, thus split into numerous thread-like segments hanging down in the water, look like roots (fig. 93).

The air absorbed by the leaves is collected in intercellular spaces which form inflated bags at the base of the branches. These help to diminish the weight of the plant and keep it floating near the surface of the water.

3. **Reproduction.**—In order to form seeds the plant throws up little stalks above the level of the water, on which two to four yellow spurred, labiate flowers are produced (fig. 93, 1).

When a pool in which Bladderworts grow dries up, they die. Before dying, they speedily make preparations to propagate their kind by producing flowers and seeds. But they can do it in another way also: the tips of the branches detach themselves from the parent plant, the leaves folding over the terminal bud; they then sink into the mud to rest there during the dry season, and sprout when the pool is filled again with water.

4. **A Carnivorous Plant.**—But the plant leads an extraordinary life not only in so far as it lives in water, but also because it feeds on animal substances. It is a carnivorous plant. To catch its prey it is provided with curious traps, which we shall now examine. If the ordinary Utricularia, which has yellow flowers and floats in quiet waters, cannot be obtained, the

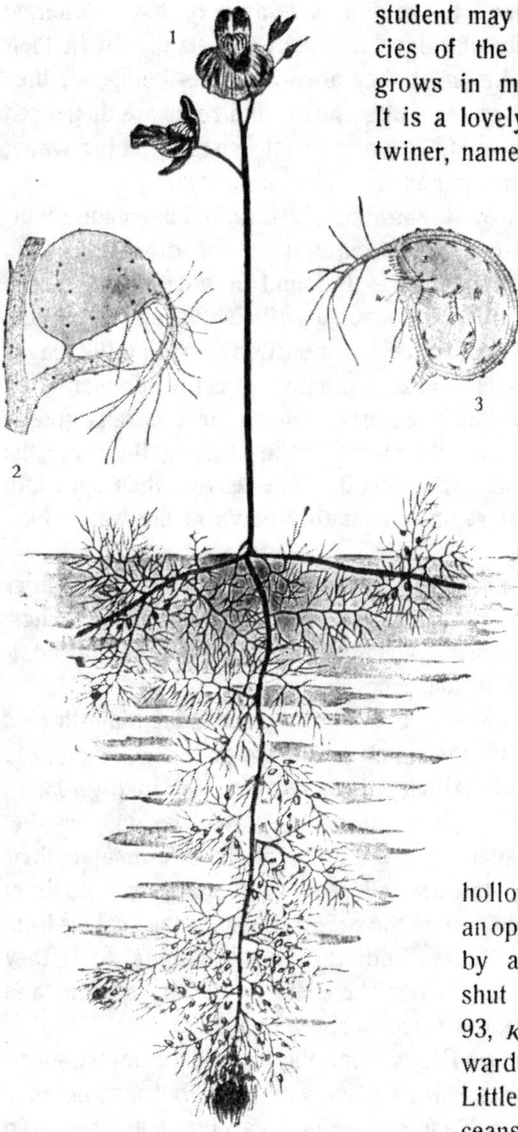

Fig. 93.—1. A flowering branch of the Bladderwort *(Utricularia stellaris)*, ½ natural size. 2. Trap as seen from without. 3. Vertical section of the trap (10 times enlarged). *K*. Valve. *W*. Abutment. Two water animals are imprisoned in bladder 3.

student may find another species of the same genus that grows in many paddy-fields. It is a lovely, blue-blossomed twiner, named *Utricularia reticulata,* with a few inconspicuous leaves, which sinks its roots into the soft mud of the rice-field. Take it out very carefully, wash away the mud, and examine it. Then you will find little knobs, here and there, of the size of a pepper corn or a little smaller. These knobs are the traps. They are hollow bladders with an opening, surrounded by a few hairs, and shut by a valve (fig. 93, *K*.) that opens towards the interior. Little snails and crustaceans that happen to seek shelter in these bladders, can easily enter, guided by special growths at the entrance of the trap, but they cannot get out again and are thus imprisoned, as the valves do not open towards the outer side. After a few days these little animals die, decay and are absorbed by the plant.

5. Many other species of Utricularia are found growing during the monsoon on rocks. Any of them presents a good type of the **ephemeral** (short-lived) **rock-plants**.

They are usually found in company with the following rock-plants: *Drosera indica*, the **Indian Sundew** (*Kan.* Pushpakāsisa, Krimināšini; *San.* Šiširapatra), with linear and glandular leaves, and *Drosera Burmannii*, with a rosette of obovate, glandular leaves belonging to the *Droseraceæ*;

Ramphicarpa longiflora, a pretty, small plant, belonging to the *Scrophularineæ*, with leaves divided into many linear segments and with large snow-white, fragrant flowers, opening in the evening;

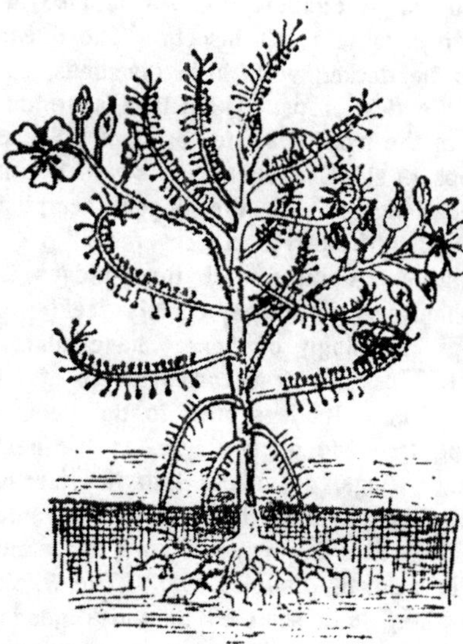

Fig. 94. — Indian Sundew (*Drosera indica*). Natural size.

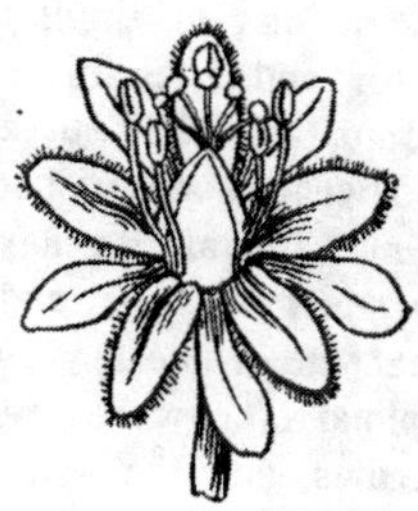

Fig. 95. — Flower of Sundew.

Eriocaulon sp. (*Kan.* Svētaširassu), a small herb with a rosette of grass-like leaves and with a white head of minute flowers (Family: *Eriocaulonaceæ*);

Burmannia cœlestis, a very small, leafless plant with two or three sky-blue three-winged flowers (Family: *Burmanniaceæ*);

Aneilema nudiflorum (*Kan.* Nelačaḷu soppu, Saṇṇaguṇḍu hullu), also a small grass-like plant, bolonging to the *Commelinaceæ*, with pretty pale-blue flowers on terminal panicles.

7*

These plants find very little food on the rocks, and have to develop their various organs within a few weeks as they cannot live when the rains are over.

The following are their adaptations for a life under such adverse conditions:—

(a) Their *seeds* are extremely *small*, they fall into the crevices of the rocks and germinate when the rains begin.

(b) The *plants* remain *small* (one to two inches high) and *grow quickly*, producing flowers and seeds in a few weeks, so that all is ended when the rains cease.

(c) The want of mineral food is compensated for, at least in some of them by *animal food*. We have seen how Utricularia catches animals and feeds on them. Drosera or Sundew (fig. 94) is also an insectivorous plant, but has other organs to catch insects. The linear leaves of this tiny plant seem to be decked with little diamonds, reflecting the light of the sun like little dew-drops. There is a sparkling and shining about them like that of the most beautiful brilliant. These sparkling diamonds are little drops of slime, a sticky mass which can be drawn into long threads when touched with the finger—secreted by numerous hairy glands with which the leaves are covered all over. What is the use of these glands? A swarm of tiny midges is dancing over the rock. One of them has left its gay society seeking a resting place. The shining drops, seemingly of honey, have allured the animal to alight on the plant. Instantly it becomes aware of its error and tries to fly away. But alas! its legs stick to the gummy liquid, and if it succeed in setting free one or two legs, in the next moment its other parts, head and wings, get again into the fearful slime, and all its desperate struggling has only the result that more glands bend over and fasten on it. This is carried to such an extent that the whole blade of the leaf occasionally doubles over. The whole animal is now covered with the slimy fluid secreted by those glands. It dies, and the soft parts of it are in course of time digested by the liquid which is slightly acid, as can be told by testing it with blue litmus paper. After the plant has thus absorbed its animal food, the leaf is unfolded, the hard skeleton of the dead animal soon dries and is carried away by the wind; and the glands shine as before till another careless fly, or a thirsty little caterpillar, or a small butterfly sits on them and runs into death. Larger insects are not caught, as they are strong enough to make their escape from the tentacles of the plant.

23. The Acanthus Family

(Acanthaceæ).

Herbs or shrubs, with decussate exstipulate leaves. Flowers zygomorphic, generally two-lipped, bracteate. Stamens inserted in the corolla, either four (didynamous) or two. Ovary superior, of two carpels. Fruit a capsule, containing a definite number of seeds and opening elastically to eject the seeds.

The Adhatoda Plant *(Adhatoda vasica).*

(*Kan.* Āḍusōge. *Mal.* Āṭalōṭakam. *Tam.* Āḍadoḍai. *Tel.* Aḍḍasaramu.)

1. **Use.**—An unattractive shrub used for fencing, easily taking root from its nodes. Cattle, even goats, do not browse on it. They dislike its smell and its taste.

2. **Stem** and **Leaves.**—Internodes pretty long and swollen above the nodes. Glabrous leaves opposite, lanceolate, acuminate, entire, secondary nerves very prominent.

The leaves in the long-stalked spikes somewhat different from those below. In many plants, *e.g. Hibiscus esculentus,* we can observe a gradual decrease in the size of the foliage leaves from below to the top. In some plants the transition is rather abrupt. And such leaves that grow from the peduncle of an inflorescence and from the axils of which flowers or branches of the inflorescence spring, are termed *bracts.* Bracts are found in the inflorescences of most plants, but they may also be absent. In the plants belonging to the Acanthus family they are always present, and here they are commonly large and conspicuous. In the axils of the opposite bracts of Adhatoda we find solitary flowers, each supported by two leaves, which are called *bracteoles.*

3. **Flowers.**—Calyx of five sepals. Corolla white, two-lipped. Upper lip of two petals, curved forward, lower lip of three petals. Æstivation imbricate. Nectar-guides consisting of red spots and streaks on the lower lip; nectary at the base of the flower: a disk round the ovary. Stamens two, epipetalous, overhanging, with their anther cells unequal in size and one slightly below the other.

The lower lip forms a nice landing place for visiting insects which when probing the tube for honey, shake the stamens and have their back powdered over with pollen.

Ovary two-celled. Style long.

Fig. 96. — *Adhatoda vasica.*

4. **Fruit.**—A bilocular capsule with two small seeds in each loculus. It opens with great force when ripe, and the stalks by which the seeds are attached to the placenta being springy, the seeds are jerked out and fall to the ground at some distance from the mother plant.

Other Acanthaceæ.

Barleria *prionitis* (*Kan.* Muḷḷugōranṭe), a thorny shrub, much branched. Thorns axillary, three-forked, representing metamorphosed branches.

Hygrophila *spinosa* (*Kan.* Kūve muḷḷu; *Tam.* Nīrmuḷḷi), a spinous annual plant growing in marshy places. The thorns represent metamorphosed branches.

Acanthus *ilicifolius* (*Kan.* Hoḷečuḷḷi), a blue-flowered plant with holly-like spinous leaves. Xerophytic like the Mangrove plant.

Justicia *procumbens* (*Kan.* Nelabēvu), a small weed with rose flowers, growing in the grass during the monsoon.

Strobilanthes, a genus represented by many species, gregariously living in the ghauts, with knotty and smooth stems, brittle like glass, flowering only after about seven years and then dying.

Thunbergia *grandiflora,* a beautiful climber with large, blue flower-bells, often grown in gardens.

Meyenia *erecta,* another garden plant with long-tubed, blue flowers.

Andrographis *paniculata* (*Kan.* Ūra-kiryātu; *Mal.* Nilavēpu) a valuable medicine plant.

Justicia gendarussa (*Kan.* Karinekki; *Tam.* Karunočči), *Asystasia violacea, Lepidagathis prostrata* (*Kan.* Ajjimuḷḷu), *Rungia parviflora* are also common.

Fig. 97.—The Til Plant *(Sesamum indicum).*

The Sesamum family *(Pedaliaceæ)* is a small family allied to the Acanthaceæ. It is represented by the **Gingily** or **Til Plant** *(Sesamum indicum),* commonly grown for the oil of its seeds.

24. The Labiate Family

(Labiatæ).

Mostly aromatic plants with square stems, decussate leaves, and bilabiate flowers. Stamens generally four: two longer and two shorter (didynamous). Style one, inserted between the lobes of the ovary, stigma bifid. Fruit of four dry, one-seeded nutlets, originating from two carpels within the bottom of the calyx.

(a) The Tumbe Plant *(Leucas aspera).*

(*Kan.* Tumbe. *Mal.* Tumpa. *Tam.* Tumbai. *San.* Rudrapushpa.)

This plant appears with the rains everywhere, near roads, ditches, hedges, and flowers as long as there is any moisture in the ground.

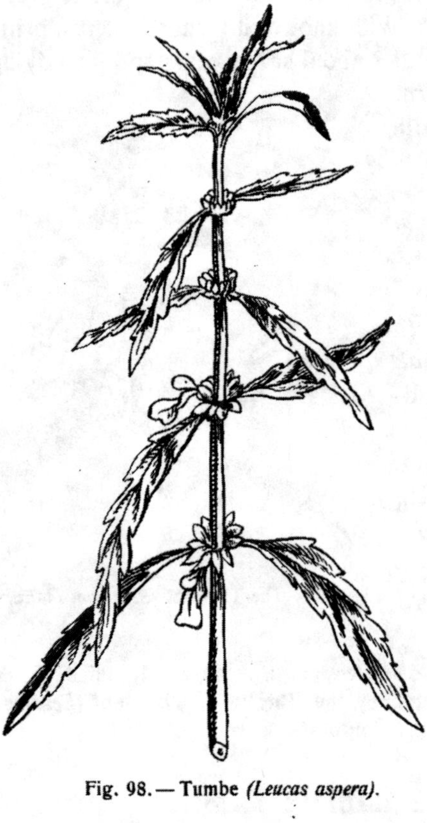

Fig. 98.—Tumbe *(Leucas aspera).*

1. **Accommodation to Conditions.**—The **Leaves** are oblong-linear, distantly serrate, slightly pubescent, and decussate, *i.e.,* they are so arranged that every pair stands crosswise over the next lower pair. So are also the many branches. This affords the advantage of the stem being equally loaded.

If we compare Tumbe plant that *grow in moist and shady places* with such as grow in *dry and sunny places,* we shall find that the former have always larger and more delicate leaves than the latter. We shall learn something from this fact.

Those plants which are growing in the shade of a tree, naturally get less sunlight. Their leaves must, therefore, be larger so as to get a greater quantity of the light that is not so intense. A small amount of the intense sunlight, which can be obtained by the small, nearly linear leaves of the specimens growing uncovered, is more than sufficient for their growth. Further, those in shady and moist places will, if plucked, fade much sooner than the other kind. Why? Since the place where they stood is always moist, they need not be economical with water and their leaves are, therefore large and tender. They lack the various means of checking the evaporation of water, such as a thick epidermis, a small surface, etc. (Contrast it with Cactus, p. 57.)

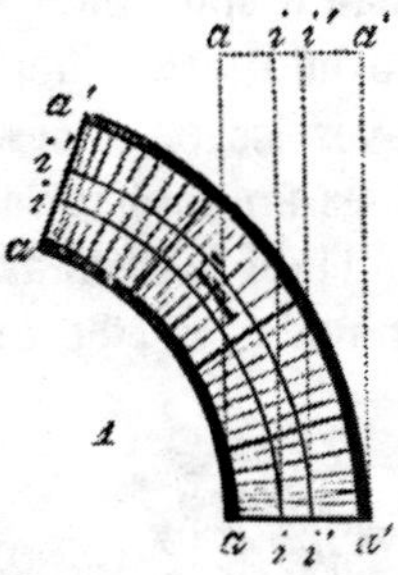

Fig. 99.—Transverse section through a cylinder, straight (with dotted lines) and bent (with thick lines). In the straight cylinder all edges are equally long; in the bent one the inner edge *(a)* is shortened and the outer one *(a')* lengthened.

The same will be found, if *plants growing on a rich and a poor soil* are compared. The difference in this case is, however, caused principally by the quantity of food the plants are able to extract from the soil; hence the root-system of those growing in the rich soil will be found to be much larger than that of plants which grow in a poor soil.

2. The **Stem** has not only to bear its own weight and that of the branches with their leaves, but it must also be able to resist the bending, twisting, and breaking influences of the wind. If the stem is bent by a rush of wind to one side, the parts on that side of the stem towards which it is bent will be pressed together, whereas the other side will be stretched by the tension exerted on it. The middle part will naturally suffer least (fig. 99). Therefore *the* **sides** *of the stem should be strongest*. Now, if the stem of the plant is cut across (fig. 100), it will at once be seen that this is really the case. There are four bundles of strong fibres at the four corners of the stem which thus becomes quadrangular.

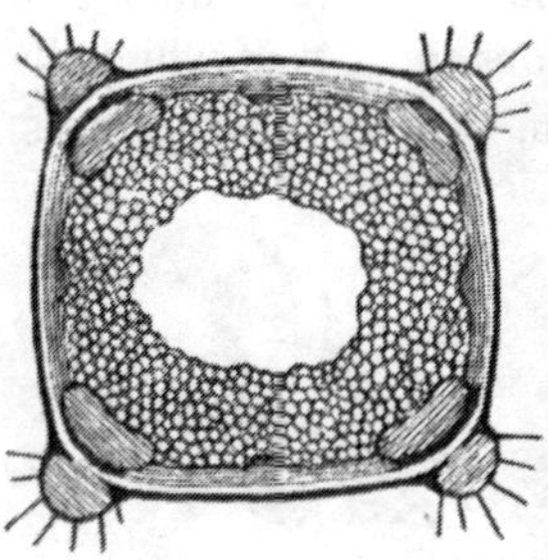

Fig. 100.—Transverse section of the stem of a Labiate plant, *Lamium album* (40 times enlarged).

And as every architect is careful to make his work as strong as possible with the least amount of material, so we see here also that the middle part which, as we have seen, has not to contribute anything towards the strength of the stem, remains *hollow* or *filled with soft pith only.*

Moreover, as a matter of fact, it is easier to break a long tube than a short one. We, therefore, find the stem of the plant divided into many short pieces by *nodes* at the parts from where the leaves issue. These nodes are solid. The pieces between the nodes are termed internodes.

3. The **Flowers** are arranged in opposite cymose fascicles on the nodes of the upper part of the stems. Calyx funnel-shaped, oblique-mouthed, with ten short teeth and as many nerves. Corolla bilabiate. The lower lip is broad, three-lobed, and forming the conspicuous part of the flower, attracting by its milk-white colour insects, which know quite well that flowers usually contain sweet honey. If you pull out one of the flowers and suck it, you will find that there really is a tiny drop of honey in each. The upper lip is much smaller and shelters, under its hairy hood or helmet, four *stamens* of which the two exterior ones are a little longer than the two interior ones.

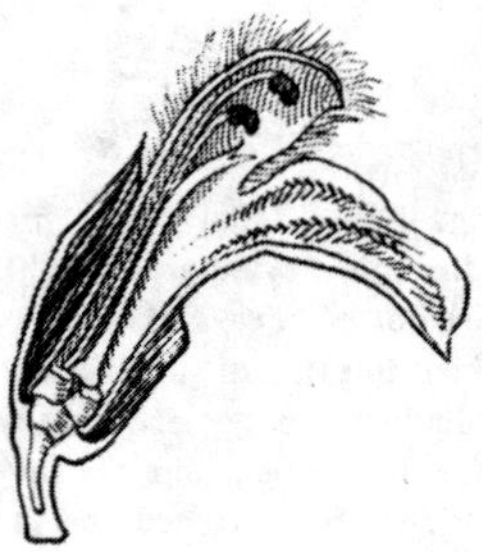

Fig. 101.—Vertical section of the flower of Tumbe *(Leucas aspera)*, only two stamens are visible.

If you now put a pencil into the throat of the flower, the stamens will slightly protrude from their sheltered place and rub themselves against the pencil. The same thing happens when a bee or other insect thrusts its proboscis into the flower-tube to fetch the honey. The stamens bend forward and deposit their pollen-

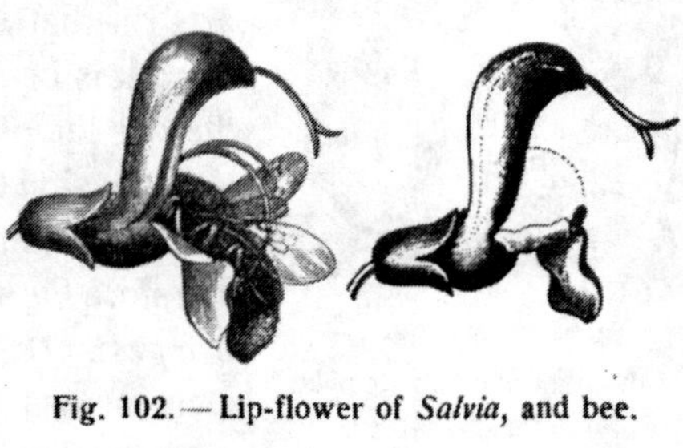

Fig. 102.—Lip-flower of *Salvia*, and bee.

grains on the back of the bee, which carries it to another flower where the pollen will fall on the two-cleft stigma, which is at the end of a long style rising from between the four-lobed ovary, and thus fertilize the ovules in the seed-box.

The stamens of *Salvia* (fig. 102) are peculiar in that their anther-lobes are widely separated by a long connective, supported by a short filament. The anther-half on the upper and longer end of the connective is developed, whereas the other half is not. The visiting insect while pushing the lower part of the connective back causes thereby the upper part to descend and deposit its pollen on its back.

Fig. 103.—The fruit of Tumbe *(Leucas aspera)*. Front part of the calyx-tube removed.

The **Fruit,** a schizocarp, is composed of four little, one-seeded nutlets at the base of the persistent calyx (fig. 103). They originate from two carpels each containing two seeds. When ripe the lobes of the carpels separate, but do not split open, and the gentlest wind can shake them out of the calyx.

(b) The Tulasi Plant *(Ocimum sanctum)*.

(Plate No. 644.)

(*Kan.* Tuḷasi. *Mal.* Šiva Tuḷasi. *Tam., Tel.* Tuḷasi. *San.* Krishṇamūla.)

This is a nice little plant which can be seen in front of most Hindu houses. It is a symbol of chastity and modesty. Its structure is very much the same as that of Tumbe. The lips of the small purple flower are, however, a little different. The lower lip is narrow and small, whereas the upper lip is four-lobed, and the stamens project outside.

The *aroma* so characteristic of the plant is due to the presence of an *ethereal oil* secreted by small glandular hairs scattered over the surface of the stems and leaves. The greater part of this oil, however, is retained by the plant in tiny casks, as it were, in which it is stored at the time of its most vigorous growth to be in readiness for use at a time when the supply of food does not keep pace with its expenditure. This happens when the plant ripens its numerous seeds, each of which requires rich food. It can be observed that the aroma of the plant is less at the time of flowering and fructification.

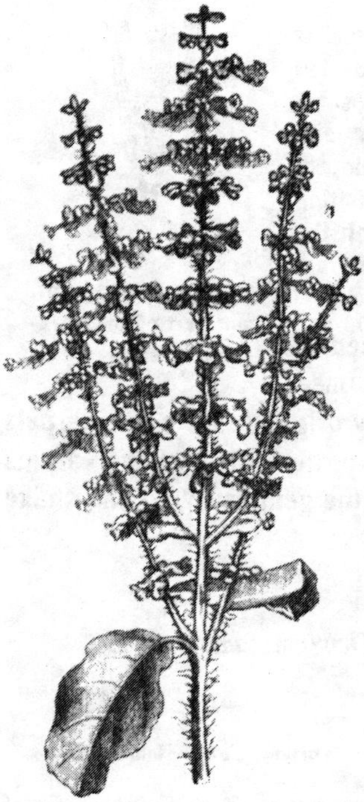

Fig. 104.— The Tulasi Plant *(Ocimum sanctum)*.

Plants growing in the shade have much less of this volatile oil than plants growing in open places. The shade compels the plant to enlarge its foliage, and thus necessitates a much larger expenditure of food stuff, so that much of it cannot be stored up for the future. Such plants do, therefore, not flower so readily as plants in open places.

In addition to the glandular hairs, the surface of the whole plant is covered by woolly hair. This coat of hairs reduces evaporation by interfering with the free circulation of air on the surface of the leaf, which we shall understand from a little experiment. Moisten two sponges of equal size and put them on the same place for drying, but wrap a piece of cloth round one of them. We shall find that the covered sponge keeps its moisture longer than the other. How does this happen? From both sponges water-vapour rises, but the vapour under the cloth cannot escape so freely as from the uncovered sponge and so the rate of evaporation is slackened. Precisely the same happens with two leaves of which one is glabrous or uncovered and the other hairy or downy. Excessive loss of moisture through the epidermis would cause the plant to wither, as it would not be in a position to make up for this loss by the sucking action of the roots.

(c) Other Labiates.

This family is but poorly represented in the tropics. They are chiefly found in the northern temperate zone, where they thrive best in a dry, sunny situation like most aromatic plants. Some common Indian species are, besides Leucas and Tulasi, **Sweet Basil** (*Ocimum basilicum; Kan.* Kāmakastūri; *Mal.* Rāma-tuḷasi), and the **Dog Tulasi** *(Ocimum canum)*.

"**Lavender** and **Salvia** for their flowers, and **Coleus** for foliage, are old-fashioned favourites in gardens; but perhaps the sweet herbs of the kitchen-garden are still better known—**Mint, Thyme, Marjoram, Rosemary, Savory.**"

25. The Verbena Family

(Verbenaceæ).

Plants resembling those belonging to the *Labiatæ*. The distinction of the two families lies in the structure of the ovary. The Verbena Family has a four-celled ovary originating also from two carpels, but with a terminal style. Fruit mostly a drupe or a berry.

The Teak Tree *(Tectona grandis)*.

(*Kan.* Tēgu, Sāgōni. *Mal.* Tēkkụ. *Tam.* Tēkku. *Tel.* Tēku. *Hin.* Sāgvān. *San.* Tēka.)

1. **Trunk and Wood.**—The Teak tree is one of the most useful timber-trees of Western India. Its wood, being fairly hard and very durable, is especially useful for shipbuilding. It also contains an oil which preserves the nails driven into it.

If the trunk is sawn through, a number of concentric circles can be seen on the section. These are found also in most other trees and are called *annual rings*, as one ring is generally formed in a year (fig. 21, page 24). They consist of alternate layers of soft and hard wood. This may be tested with the point of a knife. The soft wood is formed during the rapid and luxuriant growth of the tree in the monsoon, when the tree is covered with its enormous leaves and there is plenty of moisture, and the leaves are absorbing large quantities of carbonic acid gas from the air. The hard and dark rings of wood, however, represent the cessation of growth during the cold and dry season, when the tree drops all its leaves and stretches its bare branches towards the brazen sky. We can, therefore, estimate the age of the tree by counting its rings: for every dark, hard ring corresponds to a dry season, and every light, soft ring to a rainy season.

Trees of this class (Dicotyledons) add cylinders of wood to their trunks every year. This addition of wood is not made in the centre of the trunk, but under the bark. If the bark is stripped off at the time when the tree is growing vigorously, we always find a sticky watery fluid between bark and wood. This is the sap contained in tender cells which are dividing rapidly. The wood on the outside of the stem is lighter coloured than that in the centre, which is golden yellow when freshly cut. The layers are called *sapwood* and *heartwood* respectively. It is in the outer part of the wood, and not in the old wood in the middle of the tree, that the sap flows up from the roots to the tips of the branches to produce there, together with the food taken in by the leaves, new leaves, flowers and fruits. As new layers are formed one by one every year, the older layers cease to take any active part in the life-work of the tree, and harden. The hardest and most durable wood is the heartwood which being gradually impregnated by the waste products formed in the course of the growth of the tree, becomes denser and denser.

Fig. 105. — Portion of a net-veined leaf. *m*. Strong midrib. *n*. Side-rib.

2. The **Leaves** of the Teak tree are very large. They are *opposite,* and every pair of leaves stands crosswise to the next pair (decussate). In this way the load of the great leaves is evenly distributed, as in the Labiatæ. We can also see that the stems of young branches are quadrangular and channelled, and that they have large quadrangular pith.

The leaves are very rough on their upper side. The lower side is clothed with dense stellate, gray or tawny hairs. They also show a beautiful *network of veins* or ribs. We can guess why the ribs of these large leaves are so strong. It is, no doubt, because they are so large that they require also a strong framework to support them.

The ribs have also other functions, namely, on the one hand to carry the sap, which ascends from the roots through the trunk and the branches, to the leaf and distribute it over its whole blade, and on the other hand, to receive the products of assimilation (starch, etc.) from the green tissue of the leaf, and to carry them down through the petiole to the trunk where they are disposed of in such a way as is good for the general growth of the plant. Thus, there is a flow of sap ascending the tree to the extremest parts of its leaves, and another descending from these cells to the trunk. And for these two streams in opposite directions there are also different groups of vessels in every leaf-rib and in the leaf-stalk.

In *the dry season* the *leaves fall down*. This also is beneficial to the tree. For the leaves have such a large surface that the tree would lose too much moisture by them, and would wither and die, if they remained on the branches. (See page 52.)

Preparation for the fall of the leaves is made long before they actually fall. A fine line or ridge may be traced just below the junction of the leaf with the stem. This dark line is in reality a thin, transverse layer of cork, which when the leaves have done their work during the year, taking in stores of nourishment for the benefit of the tree, grows and so detaches the leaf from the stem. It is interesting to note also that the starch which the leaves have been making during their life-period, is not lost with them, but is transferred to the stem previous to their fall, and chiefly stored up just below the base of the leaf-stalk, so as to afford nourishment to the bud which is found in the axil of every leaf.

When they fall beneath the tree, they become leaf-mould (humus), which, in its turn, when fully decayed, restores to the soil a large proportion of the minerals taken from it by the roots of the tree. Note also when the leaves fall and decay, how the soft part between the veins rots first, leaving a beautiful skeleton of the leaf. This can be best obtained by keeping the leaves in water for a month or two.

3. **The Teak Tree a tropophilous* Plant.**—From the observations we have made above we learn that the Teak is adapted to the various conditions of the climate. The Mangrove (page 51), and the Cactus (page 57) are xerophilous in their structure, *i. e.*, they possess various contrivances to check the transpiration of water. Other plants, like the Garden Balsam (page 21), have a hygrophilous structure, characterised by large leaves with a thin epidermis to allow the water to evaporate freely.

The Teak combines the two types. It is hygrophilous at one time, *viz.*, the monsoon producing large and numerous leaves to transpire large quantities of water, and growing luxuriantly; and xerophilous at another time, *viz.*, the dry season, diminishing the process of transpiration by shedding its leaves and stopping its growth. Other trees that shed their leaves in the dry season are the Silk Cotton tree *(Bombax,* p. 15) the Pagoda tree (*Plumieria,* p. 82) and the Coral tree (*Erythrina,* p. 39).

Fig. 106.—A branch of the Teak tree *(Tectona grandis)*. Much reduced.

We call such plants tropophilous. The same is observed in most trees growing in cooler climates. There, it is during winter that plants cannot obtain water, because, though there is plenty of it in the ground, it is frozen. And as they can obtain nothing or almost nothing, they cannot spend much. Therefore they shed their leaves in the beginning of winter, and become xerophilous. Water

* From the Greek *trepo,* to turn.

becomes available for them again in spring, so they take on their green and rich foliage in summer and become hygrophilous.

4. The **Flowers** of the Teak tree are small, but clustered in large, white panicles overtopping the green foliage, and thus making the tree conspicuous at the time of flowering (compare Mango, page 27). They are white and star-like, the corolla having five or six equal lobes. The number of the stamens is the same as that of the corolla-lobes. The style is single and has a two-cleft stigma.

5. The **Ovary** is four-celled and grows, under the protection of the inflated calyx, into a very hard, bony nut covered with a fur-like coat of branched hairs.

Other Verbenas.

One of the commonest Verbenas is the **Lantana** (*Lantana aculeata; Kan.* Nātagiḍa; *Mal.* Arippu), a straggling shrub with square, prickly stems and pretty orange flowers arranged in small heads. Birds eat the fruit and spread the seed widely. It is a native of America, but has run wild nearly everywhere in India, and is a perfect curse to planters by the way in which it spreads in all directions destroying other growth. It is often used as a hedge plant. The pink or white Lantana is *L. indica.* It has few or no prickles.

The **Chaste Tree** (*Vitex negundo; Kan.* Lakki; *Mal.* Indrāṇi; *San.* Nirguṇḍi) is a tall shrub with gray foliage, covered with silvery down on the lower side and bearing small, lilac flowers in panicles. The aroma noticed in the leaves of the Labiatæ (page 107) is found also in this plant.

Stachytarpheta *indica* has long spikes of blue flowers.

Premna *integrifolia* is a strong-smelling shrub with shining, ovate leaves and greenest white flowers in large panicles.

Another very common genus of this order is the **Clerodendron** of which some species have a remarkable contrivance to exclude self-pollination.

To study this we may examine either *Clerodendron volubile,* a common garden-creeper with a white, inflated calyx and a crimson corolla, or *Clerodendron infortunatum* (*Kan.* Iṭṭēvu; *Mal.* Peragu), a handsome under-shrub with decussate, large and cordate leaves. The erect panicles of its white **flowers attract night-moths** not only by their pale colour, but also by their sweet smell which is specially strong by night. If various flowers are compared, it will be seen that some have their four stamens straight and the style bent down, whereas others have the style straight, but the stamens curled. The stamens are straight in flowers that have recently opened (fig. 107, 1), and curled in such as have already been open for one or two days (fig. 107, 2). Now, a moth that comes for nectar to a newly opened flower cannot but touch the anthers hanging on the long, horizontal stamens, with the lower side of its wings while it hovers in front of the flower thrusting its long tongue into the floral tube. Afterwards, when it goes to a flower which opened the previous night, it must touch the style of it and thus bring the pollen of the first flower to the style of the second. There is absolutely no possibility of self-pollination.

1.

2.

Fig. 107.— Flower of *Clerodendron infortunatum*. 1. Position on first evening: Stamens straight, style bent back. 2. Position on second evening: style straight and stamens curled.

As the flower of this plant attracts night-moths to avail itself of their services, so does the **fruit attract birds** by the black colour of its four drupes and the red colour of the calyx which enlarges and reddens as the fruit ripens. The birds eat the fleshy fruit and disperse the seed.

SUB-CLASS 3.—MONOCHLAMYDEÆ

Plants with a single or no floral envelope (double in some Euphorbiaceæ). Flowers frequently unisexual.

26. The Nettle and Fig Family

(Urticaceæ).

Trees, shrubs or herbs. Leaves stipulate, usually alternate. Flowers minute, monœcious or diœcious, often crowded on a fleshy body. Stamens opposite to perianth segments. Ovary superior, usually one-celled.

The Banyan Tree *(Ficus bengalensis)*.

(Plate No. 638.)

(*Kan.* Āla. *Mal.* Pēral. *Tam.* Āla. *Tel.* Maričeṭṭu. *San.* Vaṭah.)

Two peculiarities distinguish the Banyan tree: it has *(a)* very strange roots, given off by the branches and hanging down in the air, and *(b)* flowers that are hidden in globular receptacles, generally called figs.

1. There is hardly any other tree which spreads its **Roots** so wide as the Banyan tree. The Mango tree extends its roots in the ground about as far away from the trunk as the branches in the air go. The Banyan tree is not content with so much, it seeks its nourishment in an area which far exceeds the space covered by its crown.

The latter, too, is exceptionally large, as the branches spread horizontally to a great extent. The trunk could, however, not bear this load, if the long branches had no supports. It sends down *adventitious roots* here and there which enter the ground as soon as they reach it, and may become as large as, and similar to, the parent trunk. These roots, it may also be noticed, slightly bend away from the light towards the shaded interior part of the tree, thus showing their sensitiveness to light, a quality of all roots. The branching crown becomes enormously expanded, and there is formed a large hall of columns, in the shade of which there is sufficient space for a village. This power of forming roots in the air also explains a strange thing, *viz.*, Banyan trees growing on other trees and strangling them. They are not parasitic like the Loranthus on Mango trees, for they do not strike their roots into the tissue of the tree to prey on its juice. What happens is this: birds may drop a seed of the Banyan tree on another tree, where it begins to grow as an epiphytic* plant. It forms root after root. These descend the stem of the tree to

Fig. 108.—The Banyan tree *(Ficus bengalensis)*.

the ground, become stronger and stronger, and finally hug it to death. In fact, generally speaking, the Banyan leads an epiphytic life in its youth, and becomes a terrestrial plant only after some years.

2. The large, elliptic **Leaves** of the Banyan tree are downy beneath, shining above, and covered with a very thick epidermis.

* From Greek *epi*, upon, and *phyton*, a plant.

Fig. 109.—BANYAN TREE *(Ficus bengalensis)*.

1. Flowering branch 2. Longitudinal section of fig. 3. 4. 5. 6. Single pistillate flowers from within the fig.

They are full of a resinous, milky juice, as are also all the other parts of the plant. All these things work together to make the tree very *hardy* (see Mango tree, page 25).

The *leaf-buds* are protected under a sheathing cover formed by the stipules of the last leaf developed. When the leaves in the bud expand, the stipules drop to the ground and leave an annular scar on the branchlets. In cold countries, winter buds are always shut up in a case of such scales, generally glued together by a sticky substance to shelter them from the cold weather. Although resting buds sometimes have no such coverings in warm countries, we see them in this plant. They are very useful also for the Banyan tree, for they shelter the buds from withering and drying up during the dry season.

We frequently notice a *red hue on the young leaves* which indicates a very active process of breathing (see Mango, page 26).

The Banyan tree is very hardy and affords good shade. It is, therefore, often planted along roadsides. When the leaves have done their work, they fall beneath the tree. At the spot where they were joined to the stem, a distinct mark can be seen. This is a transverse layer of cells which become corky after the leaf has performed its functions, and cut the leaf off from the plant by intercepting the flow of food and water. The leaves then change their colour from green to yellow and dry up. And as they have now lost their hold of the twig, the wind or a cold night will suffice to bring them down to the ground in showers. The cork-layer which grow between the stem and the leaf now affords a protecting covering for the bare place on the stem that is left when the leaf falls off. This bare place is called the leaf-scar. (Compare Teak tree, page 111.)

3. It is often remarked by some people that Banyan trees have no **Flowers.** This mistake arises from the flowers being concealed within a fleshy receptacle, which is popularly known from the beginning as the fruit. They are called figs and are placed in pairs at the base of the leaf-stalk (fig. 109). If we cut through such a fig, we shall see that there are numerous minute flowers inside (fig. 109, 2). The round fig, then, is not the fruit of one single flower, like the guava or the pomegranate, but is

composed of a receptacle, like that of the Sunflower, with numerous flowers or fruits resting on it. The receptacle, however, is not flat, but forms a hollow ball, leaving a small opening at the top. The little flowers within the fig contain either one stamen or one pistil each, and each is sorrounded by a minute floral envelope, called perianth,* with three to five segments (fig. 109, 3-6). Hence the plant is grouped under the monochlamydeous† plants. The staminate flowers are generally placed at the top of the fig, and the pistillate ones at the bottom. The inconspicuousness of the flowers would point at pollination by the agency of the wind (see II. Part, Pollination). But the fig is *pollinated by insects.* And this is done in the following way.

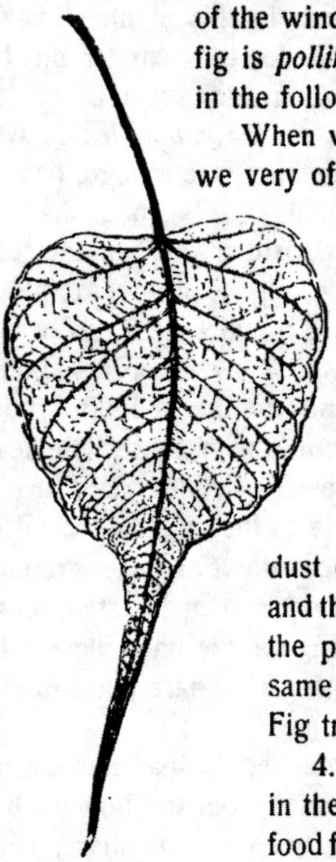

Fig. 110.—Leaf of Peepul *(Ficus religiosa).*

When we cut figs of the Banyan tree open, we very often find numerous grubs in them and sometimes also little wasps. The latter must have entered through the hole at the top of the fig. They lay their eggs in the ovules of pistillate flowers. In a short time grubs grow from these eggs and eventually become wasps again, and when they leave the fig, they cover themselves with the pollen-dust of staminate flowers near the hole, and thus, when visiting another fig, fertilize the pistillate flowers of the latter. The same happens in the fig of the Cultivated Fig tree (see fig. 112).

4. The **Figs** become scarlet and ripen in the cold weather. They are a welcome food for many birds, bats and other animals, which in their turn, disperse the seeds over a wide area.

---

* From Greek *peri,* about; and *anthos,* a flower. † From Greek *monos,* single, and *chlamys,* a cloak.

Other Fig Trees and Nettles.

The **Sacred Peepul** (*Ficus religiosa; Kan.* Araḷīmara; *Mal.* Arayāl; *Tam.* Aračamaram; *Tel.* Rāvičeṭṭu; *San.* Pippala, Açvattha) is one of the sacred trees of the Hindus. The sacred "Bo tree" of Buddha was a Peepul. Its leaves are not roundish, like those of the Banyan tree, but are drawn out in long, narrow points. When it rains we can see that the water drips from these points. And it is good that it is so. Water runs off easier from a point than it would from a blunt end, and the sooner the leaf is dry, the better for the tree. (Why?) Many trees have similarly pointed leaves, but none to such perfection as the Peepul tree. The petioles being very long, the leaves are shaken by

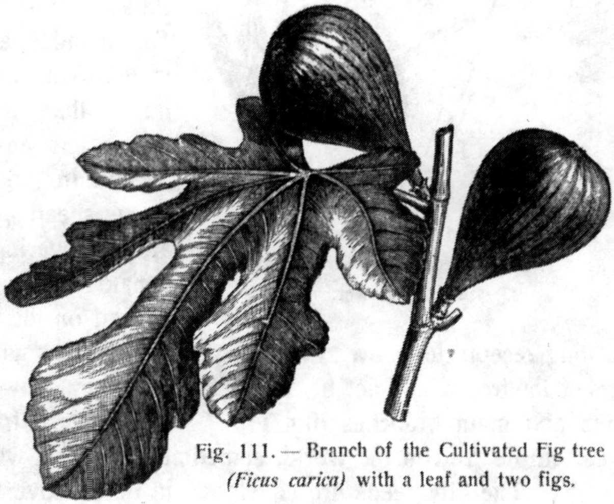

Fig. 111. — Branch of the Cultivated Fig tree (*Ficus carica*) with a leaf and two figs.

the gentlest breeze and cause a rustling noise which has given rise to many superstitious beliefs. The tree attains a very great age. The age of a famous Peepul tree at Anurādhapura, in Ceylon, was said to have been 2147 years in 1852, and must be over 2200 years now.—The fruit of the **Country Fig Tree** (*Ficus glomerata; Kan.* Attimara) is edible. It grows in dense clusters on the trunk or branches. Its leaves are often covered with galls.

There are many other species of Ficus belonging to India. The most important of the rest is the **Cultivated Fig** (*Ficus carica; Kan.* Añjūra), whose fruit forms an important part of the food of man and beast in the countries round the Mediterranean Sea, where the tree is grown abundantly, and produces a superior kind of fruit.

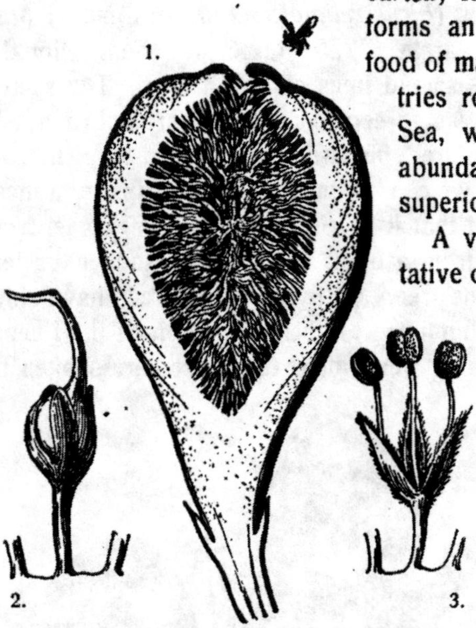

Fig. 112.—1. Vertical section of the fig of *Ficus carica*. 2. Part of receptacle with pistillate and 3. staminate flower.

A very important representative of this family is the **Jack Tree** (*Artocarpus integrifolia; Kan.* Halasu; *Mal.* Pilāvụ; *Tam.* Palāču; *Tel.* Panasa; *San.*Skandaphala). The male catkins, not much larger than a man's thumb and enveloped in bud in large deciduous sheaths, fall off after flowering; the female ones closely packed on the outside of a long receptacle, grow to be a huge fruit, to bear which the twigs of the tree would not be strong enough; they grow on the trunks and main branches (fig. 113). The immense fruit, the largest edible fruit in the world, sometimes attains a weight of sixty lbs. The oily seeds are enveloped in fleshy covers which represent the tubular perianths. The thin thread-like growths between the single fruits are sterile flowers. The tips of the perianths are hard, and appear on the outer side of the fruit like spines or tubercles. It is a peculiarity of several tropical trees (*e. g.*, the Country Fig tree and the Chocolate tree), that they have their flowers and fruits not at the ends of their twigs, but on their stems. This agrees with the fact that such trees have also a thin bark through which buds can easily break.

They need not form a thick bark as a protection against too great a loss of water by transpiration from the inner parts of the stem.

It may also be noticed that the leaves growing in tufts at the ends of the branches of the Jack tree, are not horizontally spread, but in an obliquely vertical way, thus withdrawing their surface from the most intense insolation.

The yellow wood of the Jack tree, which darkens after being cut, is used for making ornamental furniture, and its tenacious, white juice makes the best bird-lime.

Fig. 113.—A Jack tree *(Artocarpus integrifolia)*.

An ally of the Fig tree is the **Mulberry** *(Morus indica; Kan.* Rēshmikambaḷi-giḍa). The fruit of this is, like the fig, a collective fruit, with this difference, that in the Mulberry, as in the Jack fruit, the individual flowers are arranged at the outside of a common receptacle, whereas the flowers of a fig are inside the receptacle. The leaves of the tree are the food of the silk-worm. The Mulberry is largely grown both in Asia and in Southern Europe for its leaves and edible fruit.

Another representative of the Nettle family is the **Hemp Plant** (*Cannabis sativa; Kan.* Baṅgi; *Mal.* Kañčāvu; *Tam.* Paṅgi; *Tel.* Gañjāyi; *Hin.* Gāñjā). The fibre of the stem of Hemp has been used, for ages, in the manufacture of rope and cordage, canvas and sackcloth. The plant is a native of Central Asia, and has digitate leaves and diœcious flowers, the staminate flowers being on one plant and the pistillate on another. From the green parts of the plant a disagreeable smell proceeds, which can stupefy a man. On this depends also the use of the narcotic or intoxicant bhaṅgi and gāñjā, (in Arabic *hashish*), consisting of the top leaves and tender parts of the hemp plant; they are dried and either smoked in pipes, or chewed like tobacco, or made into a confection and eaten, or drunk in a dissolved form. This intoxicant renders men excitable and quarrelsome and disposed to acts of violence. It is from this latter temperament that the use and meaning of our word *assassin* (Arab. *haschâschîn* = hashish-eaters) have most probably arisen..

27. The Spurge Family

(Euphorbiaceæ).

Trees, shrubs, or herbs, usually containing a milky juice. Leaves alternate, usually stipulate. Flowers, as a rule, inconspicuous, unisexual (monœcious or diœcious). Floral envelope often wanting or simple, sometimes double. Ovary superior, usually three-celled, formed of three carpels. Fruit capsular; the carpels usually separating from a persistent axis, or drupaceous. Seeds endospermous.

The Castor-Oil Plant *(Ricinus communis).*

(*Kan.* Auḍla, Haraḷu. *Mal.* Čiṭṭāmaṇakku. *Tam.* Āmaṇakku. *Tel.* Āmudāla. *Hin.* Eraṇḍikējhāḍ.)

1. **Use.**—A large annual herb, often cultivated for the oil in its seeds, which is highly valued in medicine, and otherwise useful for lubricating machinery, dressing tanned hides, for lighting, for soap and candle-making.

2. **The Seed and its Germination.**—The seed has a hard and mottled testa, which shows at its thinner end a whitish growth—the caruncle—covering the micropyle. Break the testa and you will find the seed enclosed in a second membranous cover, the tegmen. Remove this as well, and look for the embryo: you will find the end of the hypocotyl (radicle) at that end where the micropyle and the caruncle were seen, and, above the hypocotyl, the rest of the embryo, cotyledons and plumule, not bent as in the Bean, but perfectly straight. Besides, the cotyledons are not thick but papery, and surrounded by a white oily substance, the endosperm.

Now, when we go to sow the seed and study its germination, the first sign of the growing plant is a small hook, just as it was

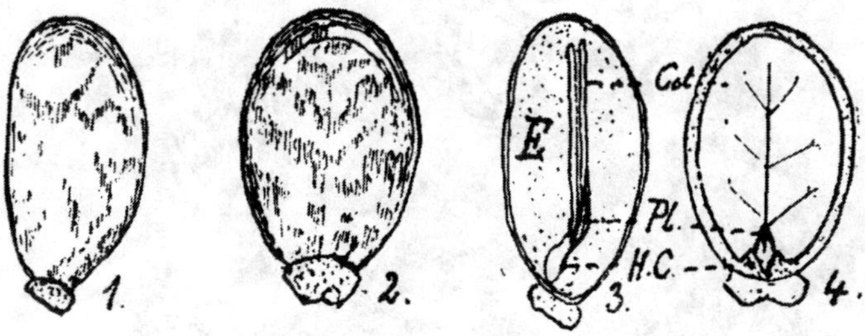

Fig. 114.—Seed of Castor-Oil plant.

in the case of the Gram seed (page 31). This hook ends at one side in a root, which soon fixes the plant in the soil, and at the other side in the seed, in which the cotyledons remain buried for some time. When they are drawn out, they are much larger and thicker than before, have a green hue and often a slimy substance on their lower sides. The testa is now empty and the endosperm is vanished. What has happened is this: the cotyledons have absorbed the fatty endosperm by their lower and outer side and, at its expense, have developed the root, enlarged the hypocotyl and also themselves; the plantlet is now established in the soil and able to obtain its nourishment from the ground and the air by means of its root and its green leaves. In a short time the plumule also grows and develops stem and foliage-leaves, which, however, differ considerably in shape from the seed-leaves.

3. **Stem and Leaves.**—Stem and leaf-stalks are covered with a bluish-white substance. Stem often channelled and hollow. Leaves alternate, peltate, stipulate. But the stipules deciduous, leaving a scar all round the stem. Blade round,

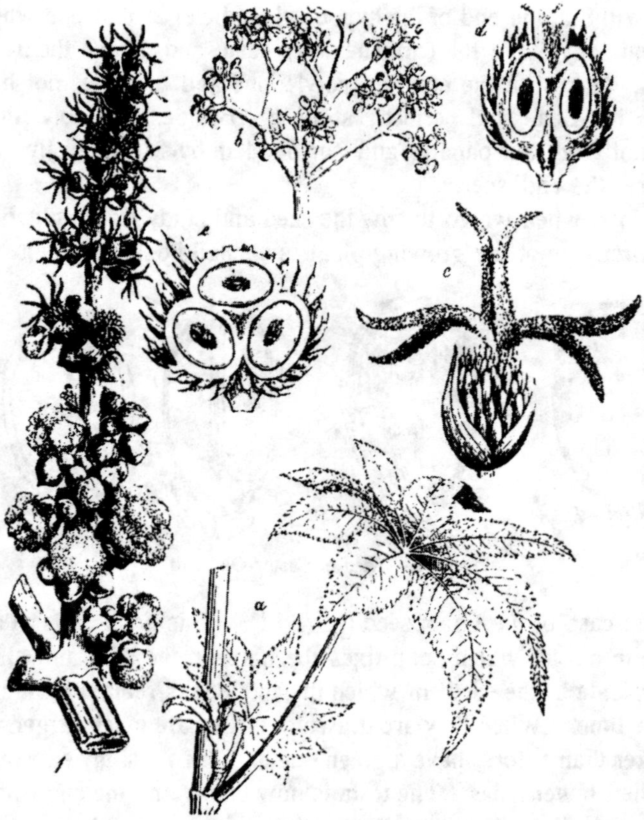

Fig. 115.— The Castor-Oil plant *(Ricinus communis)*. 1. Branch, with male and female flowers. *a.* Leaves. *b.* Stamen. *c.* Female flower. *d.* Longitudinal and *e.* transverse section of fruit.

palmately lobed, the lobes coarsely serrate and acuminate. Petiole with two large glands at the base.

4. **Flowers** in terminal panicles, composed of fascicled cymes of unisexual flowers. Usually a leafy branch springing

from the axil of the highest leaf, which, pushing aside the terminal inflorescence, continues growing in the direction of the main axis (sympodial branching). The lower flowers in the panicle only staminate, and the upper ones only pistillate (monœcious flowers). Staminate and pistillate flowers with one floral envelope of three to five sepals. Stamens much branched, and every branchlet terminating in a complete anther. Pistil of three carpels with a large prickly ovary and three long forked red stigmas. Fruit a prickly capsule with three one-seeded cells opening loculicidally with an explosion and throwing the seeds out.

Other Spurges.

The **Purging Nut** (*Jatropha curcas; Kan.* Aḍaluharaḷu), a common shrub in hedges; the **Coral Plant** *(Jatropha multifida),* a pretty garden-plant with multifid leaves and small red flowers (with calyx and corolla); the **Tapioca** (*Manihot utilissima; Kan.* Ṃarageṇasu) with an abundant deposit of starch in its swollen roots; the **Ceara Rubber Tree** *(Manihot Glaziovii)* which is cultivated for its milky juice, like the **Para Rubber Tree** *(Hevea brasiliensis);* then the **"Croton" Plants** *(Codiæum variegatum),* often cultivated in gardens on account of its leaves, which are variable in shape, mottled and variegated with red yellow and green; **Sapium** *insigne,* a middle-sized tree on rocky soil with rugged bark and horizontal branches, leafless from January to April, succulent during the rains; and **Macaranga** *indica* (*Kan.* Uppalige; *Tam.* Uttatā), an evergreen tree with peltate leaves: all these, more or less common in South India, have flowers of a similar structure.

A different type of flower we find in the Genus **Euphorbia.** Here they are arranged in heads resembling a single flower, consisting of a calyx-like involucre with four to five lobes alternating with fleshy glands, enclosing a number of male and one central female flower. The male flower consists of one anther supported by a jointed filament, the joint indicating the base of the flower, from which in allied species the perianth is developed. Of this genus we may mention *E. nivulia* (*Kan.* Bēlikaḷḷi), *E. tirucalli* (the Milk Hedge; *Kan.* Kōḍukaḷḷi), *E. pilulifera* (the

Asthma Plant; *Kan.* Nene akki soppu). *E. nivulia* is a xerophilous succulent with a milky sap, and having the general appearance of a chandelier (fig. 116). It is out of leaves during the greater part of the year, but with its persistent stipules on, which are modified into sharp spines seated on the tubercled nodes in five spiral lines up the stem. The beautiful garden plant **Poinsettia** *pulcherrima* with scarlet bracts, and the **Slipper Plant** *(Pedilanthus tithymaloides)* have similar flowers.

Fig. 116.—*Euphorbia nivulia.*

A third type of flowers is seen in the genus **Phyllanthus,** where the flowers are not united into heads, and have perianths just like the Castor plant, but they have two ovules in each cell of the ovary, whereas the Castor has only one. The **Awla Tree** *(Ph. emblica; Kan., Mal., Tam.* Nelli) is a deciduous tree, with gray bark peeling off in couch-like scales and with hard red wood showing the medullary rays very clearly on a section. The branchlets are jointed, and bear bifarious linear leaves, so that they are sometimes erroneously taken for pinnate leaves. Its fruit is an edible drupe. *Ph. niruri* (*Kan.* Nela-nelli) is a small weed found commonly in the rains.

28. The Birthwort Family

(Aristolochiaceæ).

Climbers. Leaves alternate, exstipulate. Flowers regular or zygomorphic, perianth tubular, anthers six, sessile, inserted round the base of the stigma. Ovary inferior, of six carpels. Placenta parietal. Seeds endospermous.

The Indian Birthwort *(Aristolochia indica).*

(Plate No. 241.)

(*Kan.* Īśvara-bēru. *Mal.* Perumarunnu. *Tam.* Īśuravēr. *San.* Ahigandhā.)

This is a smooth climber, growing in the thicket of jungles. The leaves are variable in shape: linear, ovate, or obovate-oblong, and have rounded or auricled bases, and their petioles serve as tendrils in support of the winding stem.

The flowers that rise in the axils in small racemes exhibit a peculiar structure. The dark green perianth is tubular, forming a globular cavity at its base and a brown tongue at its upper end. If we split the tube open, we find in the kettle-like cavity the broad end of the pistil crowned with a six-lobed stigma. Round the base of the stigma there are six sessile anthers. In fresh flowers we find bristles in the narrow floral tube pointing downwards to the cavity.

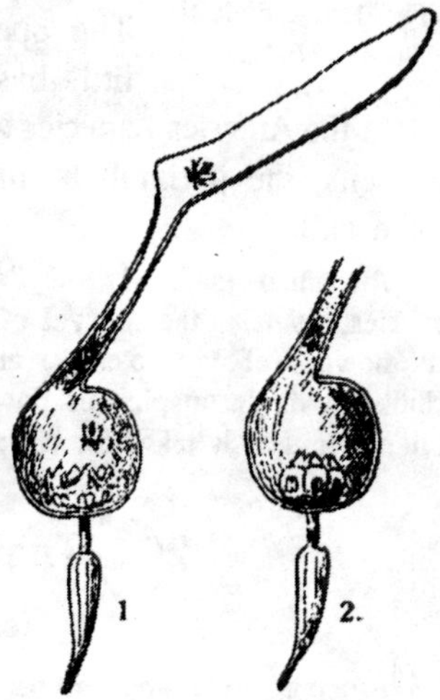

Fig. 117.—Longitudinal section of flower of *Aristolochia*. 1. Before pollination, 2. after pollination.

Such a peculiar flower can be pollinated only in a peculiar manner. Small midges attracted by the smell of the flower enter into the cavity. If they come from other (older) flowers, they bring pollen with them and brush it against the stigma, which in Aristolochia ripens before the stamens. The juicy walls of the prison in which the insects are caught give them nourishment. After about two days the stigma shrinks, but the anthers at the base of it open and let the pollen fall. The insects are powdered over and over with the pollen when they move in their little trap. Simultaneously the hairs in the narrow tube, which hitherto prevented their escape, begin to shrivel, and allow the insects a passage out. The insects now crawl out of their prison and visit another flower. The pollinated flower, however, now covers the entrance into its cavity by laying the tongue-like limb of its perianth right over it, in order to prevent insects from visiting them again.

Fig. 118. — Fruit of *Aristolochia*.

The fruit is a six-valved capsule, splitting along its partition-walls, so that the six carpels come quite free of each other (septicidal). The open fruit thus looks like a graceful little basket suspended from the climber.

In the American species *A. gigantea,* which is grown in Indian gardens, the perianth is almost large enough to form a bonnet for a child.

An allied family is the *Nyctagineæ* with two well-known garden species, namely, the **Marvel of Peru,** *Mirabilis dichotoma,* which opens its flowers at four o'clock, and **Bougainvillea** *spectabilis,* a spinous climber with inconspicuous flowers surrounded by large carmine bracts often erroneously taken for the petals.

29. The Laurel Family

(Lauraceæ).

Aromatic trees and shrubs. Leaves exstipulate, entire, gland-dotted. Flowers radial. Perianth tubular, consisting of two series of three leaves. Stamens in two or more whorls of three leaves each, filaments flattened, anthers opening by valves. Ovary of three carpels, one-celled. Fruit a drupe.

The Cinnamon Tree *(Cinnamomum zeylanicum).*

(Plate No. 632.)

(*Kan.* Dālčini. *Mal.* Karuva. *Tam.* Lavaṅgapaṭṭai. *Tel.* Lavaṅgapaṭṭa. *Hin.* Dālčin.)

The Cinnamon tree grows wild on the Western Ghats, and is cultivated in Ceylon for its bark which is a very valuable spice.

1. When we rub its **Leaves,** a fine aroma is produced, caused by a volatile oil contained in them. The same oil occurs also in other parts of the tree and chiefly in the inner part of its bark.

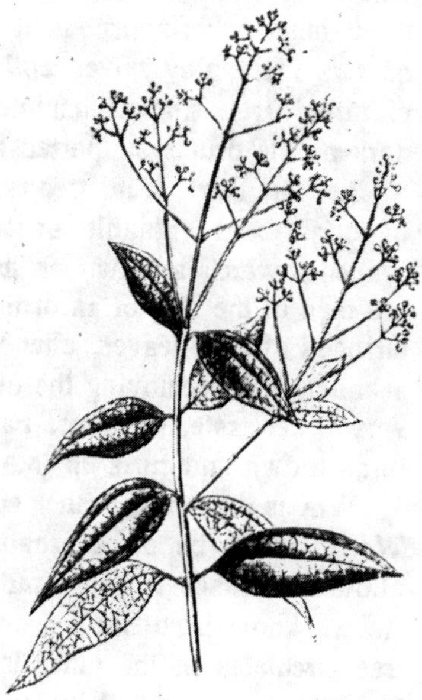

Fig. 119. — Flowering branch of the Cinnamon tree *(Cinnamomum zeylanicum)*.

The *young shoots* of the tree are often of a *dark crimson* (Plate No. 632, 2), especially in trees growing on high mountains. In the description of the Mango tree (page 26) we have already learnt that this is an indication of active breathing in the young parts of a plant. We can artificially produce the reddening of leaves by wounding them, for by doing so we increase the action of breathing, by which the plant seeks to heal the wound. (Compare the increase of breathing by which fever is accompanied.) The vigorous process of breathing increases the heat produced by the oxidation of carbon, which naturally benefits a plant growing in the cool climate of higher elevations. At the same time the red tint in the cells of young leaves may protect the chlorophyll-granules from the destroying effect of too intense light.

Fig. 120. — Flower of Cinnamon (longitudinal section, much enlarged).

Another characteristic of the opposite and leathery leaves of the Cinnamon tree are the three conspicuous parallel ribs running from the base to the tip (fig. 119).

2. The greenish **Flowers** are seated on axillary, sometimes terminal panicles. The perianth is composed of two united whorls

of three petals each. The staminal leaves are arranged in four whorls of three leaves each alternating with one another, the three outer whorls producing fertile stamens with four-celled anthers opening by valves, and the inner whorl forming a nectary of three arrow-shaped staminodes. Ovary one-celled. Fruit a dark purple drupe, supported by the enlarged perianth.

3. The **Bark** of the tree is aromatic, and has become a very important and valuable article of commerce. The bark of branches which are two or three years old is the best. They are then of the size of an ordinary cane. The branches are cut, stripped of their leaves, after which the bark is peeled off with a knife. After removing the outer part of the bark, which has a very bitter taste, the inner part of it is carefully dried, when it turns brown and curls up into little rolls.

Why is the whole branch cut when only the bark is wanted? Would it not be more economical to take away the bark and allow the branch to grow and add new bark for another time? As we know from our lesson on the Teak tree, the sap of the tree circulates in the inner layer of the bark and in the outer layer of the wood. If these are destroyed, the circulation from the root to the branches is stopped, and the branch must die. We could not possibly, therefore, expect the branch to grow after taking away its bark.

Allied Plants.

The **True Laurel,** *Laurus nobilis,* is a plant of the temperate zone. Camphor is obtained from *Cinnamomum camphora.*

To this family belongs also **Cassytha** *filiformis* (*Kan.* Bēḷu-baḷḷi; *Mal.* Ākāçavaḷḷi; *Tam.* Kottān; *Tel.* Pāčitige), a leafless, yellowish-green twiner that runs over hedges in a tangled mass. If we examine it, we find that it has no roots in the ground (hence the vernacular names!), but that there are swellings in the thread-like stem wherever it comes in contact with the plant on which it is climbing. From these swollen parts roots come out which break through the bark of the host, from which the guest sucks up its nourishment. Such plants are called *parasites.**

---

* From Greek *para,* beside, and *sitos,* food.

An allied family is the Nutmeg family *(Myristicaceæ)*, of which the **Nutmeg Tree** (*Myristica fragrans; Kan.* Jāyi) is generally known. Its well-known seed has a red lacerate aril growing from the hilum, and an oily, ruminate endosperm

30. The Mistletoe Family

(Loranthaceæ).

Parasitic shrubs. Leaves often fleshy. Petals four to eight, free or united. Stamens as many and opposite to the petals. Ovary inferior, one-celled and one-seeded. Fruit a berry or drupe.

The Loranthus *(Loranthus longiflorus)*.

(*Kan.* Bandaṇige, Baḍanike. *Mal.* Pulluṇṇi.)

1. **Parasites and Epiphytes.**—Certain plants grow on trees, having leaves and flowers different from those of their hosts, growing out of their branches like twigs, as if they were grafted on them. The vernacular names, generally given to these plants, mean what in English is called *parasite.** Such plants insert their roots into the stems of the plants on which they settle and derive their nourishment from the juice of the host.

Parasites must be clearly distinguished from *epiphytes.*† The latter settle also on branches of other trees, but have their roots only on the surface of their bark providing their food for themselves. Such plants are, *e.g., Philodendron sp., Ficus sp.,* some Orchids, many Mosses and some Ferns, such as the gigantic *Polypodium quercifolium* with its oak-leaved scale over the bearded rhizome.

2. **The Loranthus Plant a Parasite.**—Now, if you examine the roots of a Loranthus plant, you will find that it penetrates right into the interior part of its host. It must, therefore, be considered as a parasite.

But how did the Loranthus plant obtain its lofty place on the branch of a tree? Certain birds are fond of its juicy berries,

---

* See footnote on page 130. † From Greek *epi,* upon, and *phyton,* a plant.

and as these berries have a clammy substance, the seeds often stick to the beaks of the birds and are thus carried away by them. When they happen to rub their beaks on a branch, the seed sticks to it and germinates there eventually, the tiny root

Fig. 121.—Epiphytes on a tree: *Philodendron cannæfolium*, and *Ficus sp.*

making its way into the soft bark until it comes to the hard wood-cylinder in the interior of the branch through which it cannot pierce. Side-roots are then developed which creep along the wood and below the bark sending out additional vertical roots into the interior (fig. 122). As the host grows, these roots are covered by new layers of wood, so that, in course of time, the roots of the parasite appear to be in the very centre of the branch and to be fused with the wood tissue of the host. Now, the woody tissue of a tree is the channel for the ascending sap (see II. Part, Stem). And it is, therefore, the ascending water and its mineral ingredients which the Loranthus takes from the host. Other nourishing substances, such as sugar, starch, albumen, are conducted in vessels that are in the inner bark of the tree. The Loranthus does not feed upon them, but prepares them by itself. And for this purpose it requires green leaves like other plants. The Loranthus may, therefore, be called a *Water-Parasite*.

Fig. 122.—A young Mistletoe *(Viscum album)*, the roots being laid bare.

Notwithstanding this fact, the Loranthus destroys the host. As we have seen, it interferes with the flow of sap in the woody tissue of the host. Accordingly the host is weakened, and not infrequently the branch is killed by the intruder. If the parasitic bush is only cut off, the roots lying in the tissue of the host will sprout again. The only remedy to save the tree is to cut off the branch below the place where it is attacked.

Many other parasites feed not only on the ascending sap of their hosts, but grow entirely at their expense, taking water, starch, nitrogenous and every other kind of food they require from their host. Such plants need no longer develop any organs to absorb raw food from the air and, therefore dispense with leaves. In Cassytha (page 130) we have met with such a *holoparasite*.*

. On the other hand there is still another category of parasites which we may call *hemiparasites*† of which the sandalwood tree, *Santalum album* (*Kan*. Šrigandha mara; *Mal.* Čandana maram), is an example. This tree has roots of its own in the ground and develops green foliage like ordinary trees. Yet it is a parasite. When its roots come into contact with roots of other plants, they attack them and suck food from them. This explains a curious fact which is observed now and

* From Greek *holos*, whole. † From Greek *hemi*, half.

then: the growth of a Sandalwood tree is affected when a tree in its vicinity is felled, which shows that the Sandalwood tree has lost a source of its nourishment.

3. **Leaves and Flowers.**—*Loranthus longiflorus* is the commonest of the numerous Indian species of this genus, and it is frequently found on Mango trees. Its leaves are leathery, opposite, and extremely variable in shape, the side-nerves being obscure. Flowers long, in one-sided racemes. Calyx short and truncate. Corolla curved, long-tubed; tube scarlet, or orange, or pink; lobes five, linear reflexed, green or yellow. Stamens inserted on the petals, opposite to them, with linear anthers. Ovary inferior, one-celled with one ovule. Fruit a pulpy berry, crowned by the calyx.

Loranthus lageniferus is a species that sends out aerial roots which creep along the branches of the foster-plant and form suckers. The flowers of this species are supported by coloured, bell-shaped involucres.

31. The Pepper Family

(Piperaceæ).

Aromatic herbs or shrubs. Leaves entire. Flowers achlamydeous, minute, in catkin-like spikes. Ovary one-celled. Seeds endospermous.

The Pepper Vine *(Piper nigrum)*.

(*Kan.* Oḷḷēmeṇasina-baḷḷi. *Mal.* Kuru-muḻagụ. *Tam.* Miḷagu. *Tel.* Šāvyamu. *Hind.* Kālā-miriči.)

1. This is a large **Climber** requiring the support of other trees. It climbs, however, not like the Bean by winding round its supporter, or like the Cucumber by using tendrils for this purpose, but with the help of small clinging roots which grow from the swollen nodes of the slender zigzag stems, winding round their support like little cords, and contract after some time so that the climber is drawn close to the tree, and fastened to it, as with a thousand little fingers. They do not penetrate into the tree, and hence the Pepper vine is not a parasite. Besides these adventitious roots, the Pepper vines have, like

ordinary plants, proper roots in the ground. If these are cut, the plant withers unless, on its way up the tree towards the light, the plant has found nourishing earth in the crevices and holes of the tree, in which case, the clinging roots may become feeding roots, changing their function.

2. The **Leaves** are alternate, stipulate, ovate, entire, and have three or more basal nerves, which stand out prominently at the lower surface.

The wood of many climbing plants is not arranged in concentrical continuous cylindrical tubes, as it is in the case of most other plants (see Teak tree, page 109). In the Pepper plant the woody tissue of its stem is separated by broad medullary rays into wedge-shaped masses.

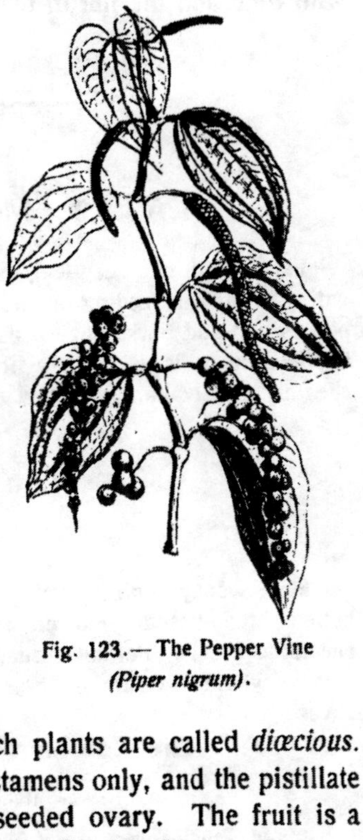

Fig. 123.—The Pepper Vine *(Piper nigrum)*.

3. **Flower and Fruit.**—When the plant has ascended the tree, and so reached a point where it gets more light, its stem leaves the trunk of the tree, and produces no more adventitious roots, but forms flower-buds opposite the leaves. The flowers are arranged on leaf-opposed (see Grape Vine, p. 22), hanging spikes, and are very small, without any regular calyx or corolla. They are unisexual, *i. e.*, bear either stamens or pistils, but not both together. Besides, one plant has only staminate, and another only pistillate flowers. Such plants are called *diœcious*. The staminate flowers have two stamens only, and the pistillate ones have a one-celled and one-seeded ovary. The fruit is a berry which becomes red when ripe. Seed endospermous.

4. **Use.**—The plant is grown for the fruit, which is used as a condiment. Black pepper is the unripe, dried berries; white pepper, the same allowed to ripen, with the pulpy coat removed.

The pungent taste and smell of the pepper corn is due to an aromatic oil. The same substance is also noticeable in its leaves, but to a lesser degree.

The propagation of the Pepper vine is effected by means of mature branches. These are layered, *i. e.,* bent down into the ground, and when they take root, they are severed from the parent vine, planted out in shade, and trailed on to trees.

5. A near ally of the Pepper vine is the **Betel-Leaf Pepper** (*Piper Betle; Kan.* Vīḷyada-baḷḷi; *Mal.* Tāmbūlam; *Tam.* Vettilaikkoḍi; *Tel.* Tamalapāku; *Hin.* Pān). Its leaves are chewed with lime and the nut of the Areca palm.

CLASS 2.—MONOCOTYLEDONS

Plants with one seed-leaf, usually remaining enclosed in the germinating seed, and feeding on the endosperm. Stems not separable into pith, wood and bark, but consisting of fibro-vascular bundles, scattered in the cellular tissue, with a firmly adherent rind outside. Leaves on sheaths, usually parallel-veined. Floral parts generally in sets of three.

32. The Palm Family

(Palmæ).

Stem woody, unbranched. Leaves pinnately or palmately divided, petiole broad-based. Flowers radial, in panicles or spikes, with large sheathing bracts. Perianth usually six-leaved in two whorls, stamens six in two whorls. Ovary of three carpels. Fruit a drupe with a fibrous covering.

The Cocoanut Palm *(Cocos nucifera).*

(Plate No. 637.)

(*Kan.* Teṅgina-mara. *Mal.* Tēṅa. *Tam.* Teṅgu. *Tel.* Teṅkāyičeṭṭu. *Hin.* Nāralkejhāḍ. *San.* Tṛiṇarāja.)

The Cocoanut Palm is a tree found only in tropical countries, and there grows best near the seacoast.

Fig. 124.—The Cocoanut Palm *(Cocos nucifera)*.

1. Its slender, cylindrical **Stem,** and the tuft of leaves with which it is crowned, is so different from the appearance of other trees that every one at once understands that the Cocoanut Palm belongs to a class of plants quite different from that which most other trees belong to.

2. **Monocotyledons and Dicotyledons compared.**—A general comparison of the Cocoanut tree with, for instance, a Mango tree will make the characteristic features of the new class, called Monocotyledons, distinct and clear. Beginning with the *root,* we find that the Mango tree has a thick and stem-like taproot with numerous side-roots, whereas the root of the Cocoanut Palm consists of many similar, thread-like or fibrous roots.

The *trunk* of the Mango tree is stout, grows thicker and thicker as it grows older, and is, at a certain height, divided into many branches. The stem of the Palm tree is slender, does not increase in girth as it grows older, and never branches. The latter fact explains why the stem need not grow in girth, for it has not to bear such a great load as the Mango tree. An examination of the section of the stem will show that there are no annual zones in the wood-tissue, but that the substance of the stem is like a bundle of sticks closely bound together (fig. 125). The outside of the trunk is not covered with *bark,* but consists only of the very hard outer layer of the wood itself. Young stems of monocotyledons have an epidermis, *i. e.,* a thin cellular coating like that of leaves. The absence of the bark proper suggests that there is not such a thing as the cambium ring in the stem of the Palm tree. The sap circulates in the fibro-vascular bundles distributed throughout the stem. The cellular tissue—the pith—of the inner part is looser, and contains fewer vascular bundles than that of the outer part, which, with its densely packed ring of bundles, protects the inner part from damage, evaporation and changes of temperature. Many plants of this class, *e. g.,* the Bamboo and other Grasses, have even hollow stems.

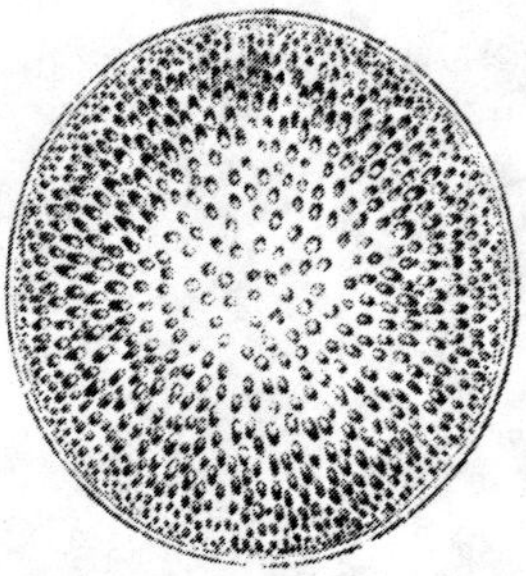

Fig. 125.—Transverse section of a monocotyledonous stem.

The name "monocotyledons," by which this new class is known, and to which the Palms, Lilies, Grasses and Orchids belong, refers to another peculiarity of theirs, namely to the fact that their *seeds* have only one leaf, the cotyledon, whereas the plants that fall under the class of dicotyledons have two seed-leaves. And this cotyledon remains enclosed in the germinating seed, as an organ for the absorption of the endosperm stored in it.

There are a few more points that characterise the monocotyledons among which we will mention only these:—

The petioles of their *leaves* have either sheaths or broad bases, and their blades are generally parallel-veined. The *floral leaves* are mostly in sets of three, whereas the dicotyledons have net-veined leaves and the floral parts in sets of four or five.

3. **Leaves.**—We have seen that in the Cocoanut Palm a crown of mighty, feathery leaves waves on a slender stem, which can reach a very considerable height; and the wind is, therefore, able to exercise its full force on the leaves. These are exceptionally large, sometimes sixteen feet long, and if their blades were entire, as they are, indeed, in their undeveloped, folded bud-state, the wind would certainly tear them into pieces or up-root the whole tree. But the Cocoanut Palm can withstand the strongest storm, as its huge leaves are slit into segments by the rupture of the tissue at the edges of the folds, forming *pinnate, i.e.,* feathery leaves, which let the wind pass between the pinnæ, and so lessen its pressure. Besides, the leaves are covered with such a *strong and hard epidermis* that no vehemence of the lashing tropical rains can do them any harm.

The tuft at the end of the stem contains twelve to twenty-four leaves. As generally *every month one leaf is produced and one is dropped,* the number of them does not increase. The base of the leaf-stalk is broad and stem-clasping, and the fallen leaves leave a very distinct scar on the stem, by counting which the age of the tree may be estimated.

4. **Cultivation.**—Cocoanut trees are commonly planted in deep pits when they are one or two years old. It is several years before the tree grows to any height. During these first years the trunk is formed till it attains its ordinary width. At the same

time, the leaves enlarge one by one until they have approximately the size of those of full-grown trees. Numerous fibrous roots are being produced to fix the tree firmly in the soil. When it has thus gradually attained the size of a full-grown tree, as regards the girth of the trunk and the tuft of leaves, it at last begins to raise itself, and henceforth ceases to grow in width. By planting it in a pit, a strong hold in the soil is secured. Later on, adventitious roots spring from the trunk higher up, holding the tree like the ropes of a flag-staff. The pit is filled up and disappears.

5. **Flowers.**—Out of the axils of the leaves spring the much-branched inflorescences which are at first protected by a huge spathe (fig. 126, 4). The spathe is torn open in longitudinal lines by the swelling flowers within. It remains a long time at the base of the gigantic inflorescence, which is a fleshy panicle divided into numerous drooping spikes, crowded with unisexual flowers. Such a form of inflorescence, in which the flowers are closely arranged round a fleshy axis and the whole surrounded by a large leaf (spathe), is termed *spadix*. It is found in the Palmæ, Aroideæ and Pandanaceæ, which families are, therefore, sometimes grouped as Spadicifloræ. The rapid development of the spadix is only made possible by a rich flow of sap from the trunk. This explains the obtaining of palm-wine by cutting the spadix.

The flowers are unisexual, *i. e.,* both sexes grow on the same plant. They are placed on the branches of the spadix in such a way that the *male flowers greatly outnumber the female ones,* and also so that the female flowers are always near the base of the panicle and the male ones at the end. It is good that the *female flowers are situated at the lower end of the panicles,* for, if the heavy nuts were suspended by a long stalk, they could easily be detached by the wind.—The staminate flowers, which can be picked up in large numbers at the foot of every fruit-bearing Cocoanut tree, as they drop after flowering, consist of three smaller sepals and three larger petals, all being horny and straw-coloured. The six stamens are also arranged in two whorls of three and three. The pistillate flowers are much larger than the staminate ones. They consist of six imbricating petals and

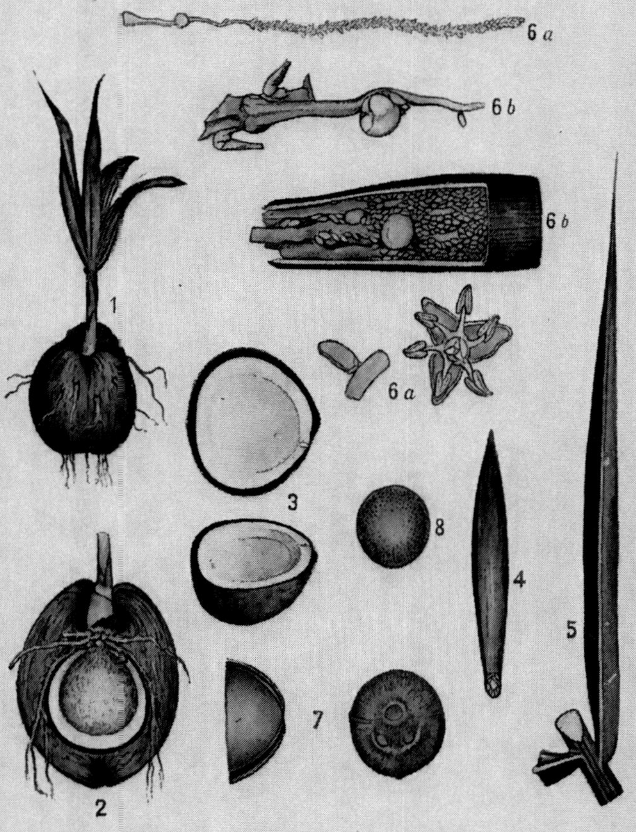

Fig. 126.—Cocoanut *(Cocos nucifera)*.

1. Young plant. 2. Vertical section of germinating nut. 3. Ripe cocoanut cut open. 4. Spathe. 5. Pinna 6. Flowers, *a*. male, *b*. female. 7. Upper part of nut-shell. 8. Cross-section of stem.

a three-celled pistil, but the petals are broad, and increase with the growth of the egg-shaped ovary, forming a large, cup-shaped base for the ripe nut.

6. The **Fruit** is a drupe. Its *covering* is twofold,—a fibrous mass outside, the exocarp, and a shell as hard as stone in the interior, the endocarp. Break the latter and you will get the kernel, which is the *seed,* formed like a hollow ball, and contaning the minute embryo in its pulp and a milky substance in its cavity.

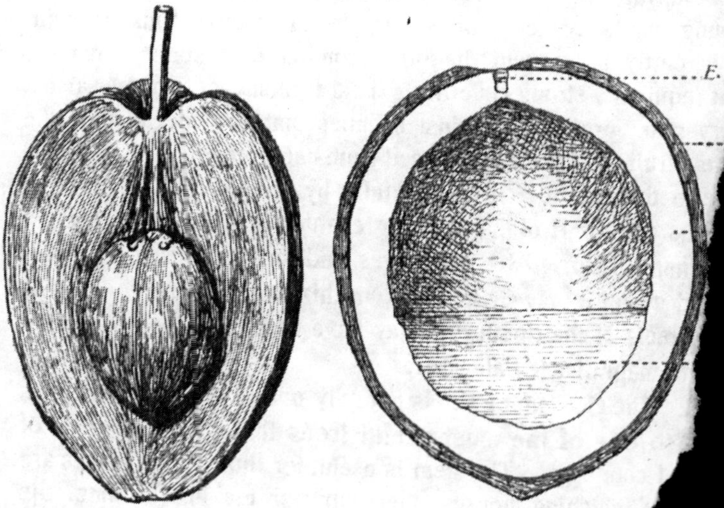

Fig. 127. — The Cocoanut with part of the fibrous covering removed. To the right the shell or endocarp *(S.)* opened, showing the kernel *(K.),* the embryo *(E.)* and the milk *(M.).*

The pulp and the milk, the endosperm of the seed, form the first nourishment of the seedling.

But the very tender *seed-bud* is evidently too weak to push itself through the hard shell. Hence that part of the shell which is immediately over the embryo or germ is so thin that it can easily be pierced by the tender sprout. Of the three carpels, of which the ovary is composed, only one develops and forms a germ, and we see the traces of the other two in the two other hard eyes which every cocoanut possesses. Moreover, the germ, and with it these holes in the shell, are always at that end of the nut where it is fastened to the stalk. The fibrous covering is least developed at this spot, and can be easily broken through by the seedling.

When the seed is allowed to germinate, the cotyledon enlarges and gradually fills the whole cavity of the seed, forming a spongy globular white body, which in course of time, eats away the kernel. The plant nourished thereby forms a bulbous body outside the shell, from which roots and leaves grow.

The *fatty oil contained in the kernel* is the chief food of the voung plant. Mixed with water, the oil soon becomes rancid. t is partly to prevent the oil becoming thus spoiled that the ut requires a strong covering. This furnishes, of course, also a ry good protection against enemies that are covetous of the eet fruit, and protects the seed from harm when it falls from the to the ground. If the nut falls by chance into the sea, the ous outer part of the covering enables it to float, and the nut then be carried by the waves and sea-currents to a distant d where it may strike root. In this way the lonely, deserted reefs of the South Sea may have come into the possession is magnificent Palm tree.

The Cocoanut Palm is not only one of the most beautiful, lso **one of the most useful trees** that adorn the coast of l countries. The stem is useful for timber; the leaves are r thatching houses, their ribs for making brooms; the ud of the young plants furnishes a palatable vegetable; apping the stalk of the inflorescence a juice is obtained, which by fermentation palm-wine (toddy) is made, and ich, if unfermented, yields a good sugar when boiled down; middle part of the covering of the fruit yields a very useful fibre, out of which ropes are made, which possess a high wer of resistance to the action of water; out of the hard shell ey make drinking vessels, spoons, *etc.*; the kernel has a delicious taste and forms part of the daily food of the people; an oil of good quality is obtained from the kernel; the refuse forms a food for cattle; the milk of the fruit is a delicious beverage: there is hardly any part of this tree which is not of some use to man.

8. **Enemies.**—Among the animals which destroy the tree and its fruits may be mentioned the rat, which bites a hole into the nut in order to get at the kernel, and the Goliath beetle *(Oryctes rhinoceros)*, which damages the trees by cutting large holes in them through the young leaf-shoots. When the leaves open, signs of the beetle's work are shown.

Other Palms.

The family to which the Cocoanut Palm belongs, the *Palmæ*, is essentially a tropical one. The unbranched trunks, marked with the scars of the leaf-stalks, and their terminal crown of noble, evergreen leaves, are characteristic of the order. So are the unisexual flowers, thickly arranged in panicles or spikes within a protecting spathe. Various species of this useful order are commonly known, as they grow everywhere in the tropics.

Perhaps the commonest in Southern India is the **Palmyra Palm** (*Borassus flabelliformis; Kan.* Tāḷi; *Mal.* Tālam; *Tam.* Panai; *Tel.* Tāṭi; *Hin.* Tāḍ—Plate No. 640). This tree chiefly grows on the slopes from the cultivated valleys to the plateaus above, or on sandy plains near the coast.

The leaves are fan-shaped, often with a spiral twist, their petioles being serrated and spinous on the edges. The flowers are diœcious. The inflorescence of the male tree consists of several three-forked spikes, supported by a spathe, each fork being about one foot long. The spikes contain hundreds of minute flowers arranged in dense cymes, each of more than twelve flowers, covered under imbricated scales. The top flower appears from under the scale, and falls off after a day, making room for the next lower one. The small flower has three whitish petals with brown streaks and six yellow stamens. The spike of the female tree is about one foot and a half in length, each flower being wrapped up in half a dozen petals, and its size being that of a cherry. The full-grown fruit is dark-brown, and half the size of a cocoanut, with very tough fibres. There are three seeds inside, consisting each of a jelly-like, hollow kernel—the endosperm—with the germ or embryo at the end.

Fig. 128. — An Indian jungle with various Palm trees.

The spadices of both, male and female trees, are cut, and the sap which flows out of the wound is drunk as toddy, or made into jaggery. The toddy intended for jaggery is drawn in lime-coated pots, then boiled, and thus converted into jaggery.

The trunk of the tree is used for rafters. The fruit can be eaten. The leaves are used for many purposes, like those of the Cocoanut Palm.

Other Palms are the majestic **Talipot** or **Fan Palm** (*Corypha umbraculifera; Kan.* Śritāḷi), which forms a huge, terminal inflorescence once in its life, and dies after the seeds ripen; the **Areca Palm** (*Areca catechu; Kan.* Aḍike; *San.* Tambūlam), the most slender and elegant of Indian Palms, "raising its graceful stem and feathery crown like an arrow shot down from heaven" *(Hooker).* The nut is eaten with betel leaves.

From the stem of the **Malabar Sago Palm** (*Caryota urens; Kan.* Baini; *Hin.* Īnd) a sago is obtained. This is the starch stored up in the soft cells of the stem, preparatory to the production of flowers and seeds before the tree dies. It is also a very lofty and noble Palm, the great hanging clusters of flowers and fruits being very noticeable, and resembling a huge horse-tail. The leaf-stalk makes a fair fishing rod, the fibre of the spathe a good line. Toddy is also extracted from the peduncle of the inflorescence, in the same way as from the Cocoanut and Palmyra.

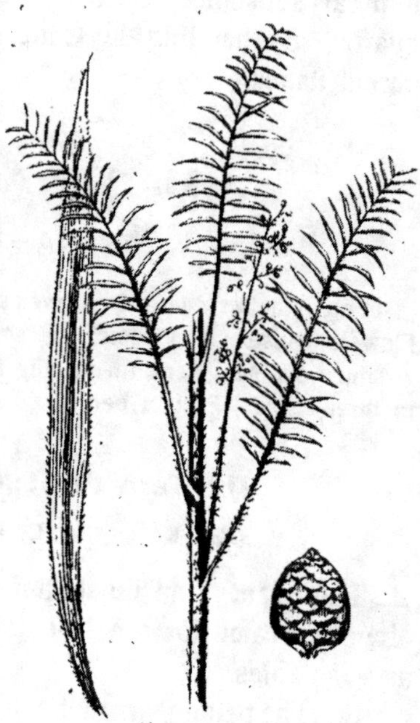

Fig. 129. — The Rattan Cane Palm (*Calamus rotang*).

What the Cocoanut Palm is to India, the **Date Palm** (*Phœnix dactylifera; Kan.* Karjūra) is to Arabia. Its fruits, that ripen in August, come to us at Christmas time.

The **Wild Date Palm** (*Phœnix sylvestris; Kan.* Īčala) grows in many parts of India. Its leaves are used for mats, and the fleshy axis of the inflorescence yields a kind of toddy.

Another Palm, fairly common in our jungles, is the **Rattan Cane Palm** (*Calamus rotang; Kan.* Betta). The midrib of the pinnate leaf of some species is produced in a very long slender drooping flagellum, resembling the slender lash of a whip. This is armed with recurved thorns on the under-side. The flagella fasten the leafy shoot of the Rattan stem to the branches of jungle trees. When the tops of the latter are reached, the leafless part of the stem glides down, as the shoot grows, and hangs from the branches of the trees in huge slings. The stem of the Rattan can thus reach the enormous length of more than seven hundred feet.—The Cane is split and manufactured into numerous articles of utility. The shining outer coating of the Rattan Cane is a secretion from the plant, and consists mostly of a mineral substance, called silex, which is nearly the same as quartz. By bending the Cane it comes off in little, thin, transparent flakes.

33. The Arum Family

(Aroideæ).

Perennial acrid herbs. Leaves alternate, net-veined, often variegated. Flowers unisexual, monœcious, sessile, small, arranged on a spadix within a spathe. Perianth usually nil. Stamens two to six. Ovary one to three-celled. Fruit a berry.

The Taro Plant *(Colocasia antiquorum).*

(*Kan.* Kesu. *Mal.* Čẽmpu̧. *Tam.* Šimaikiḷaṅgu.)

This plant is an inhabitant of shady and moist places. It is often cultivated for the leaf-stalks and tubers, which are eaten as vegetables.

1. The peltate, arrow-head **Leaves** arise, not from a stem, but from a truncated tuber, and are, therefore, called radical. They are *large and glabrous* (having no hairs). As they grow during

the monsoon, and in swampy soils, there is no danger of their being dried up. On the contrary, the high percentage of moisture in the air at that time tends to check the action of transpiration. To supplement this vital process, and to assist in its growth, the *plant is enabled to let the water pass* out *in drops from a minute pore at the tip of its leaves* (fig. 130), to which free canals, in the substance of the leaves, converge. In this way, room is made for new food-substances to be brought up by the roots. These drops can be noticed especially when the temperature is low, and the air can, therefore, not hold much water-vapour.

Fig. 130.— Leaf of *Colocasia* giving out a drop of water.

When it rains, the leaves do not become wet; the water runs off in silvery drops, as it would from a duck's back. This is due to a *wax-like coat* spread over the surface of the leaves. If the water wets the leaves, it will hinder the growth of the plant (see Lotus plant, page 2).

Cattle are careful to avoid feeding on them; for though they taste sweet at first, they leave a very acrid and disagreeable taste afterwards, which is due to the presence of certain salts in the leaves and stalks.

2. The **Flowers** are rarely seen in the cultivated kinds of Colocasia. But they can easily be obtained from the wild variety, as well as from *Caladium bicolor*, an allied plant which is grown in gardens for its variegated leaves. In the latter, the flower appears with the first leaves soon after the first rains. What is generally called the flower, is, however, not a single flower but an inflorescence—*the spadix*—consisting of a large, hood-like bract, called the *spathe,* and a fleshy spike of numerous small, unisexual flowers, so arranged that those at the bottom of the spike are pistillate and those at the top staminate, intercepted by some abortive pistils in the middle. The staminate flowers consist of sessile anthers only each opening by a minute pore at the top, and the pistillate flowers of pistils only all closely packed together on the spadix.

The spadix of Colocasia appears at the end of the rainy season. It differs from that of Caladium in having the spike prolonged beyond the stamens into an acute, yellowish club which bears no flowers, and serves as a means of *attracting insects* on which the plant depends for the fertilization of its ovules (figs. 131, 132).

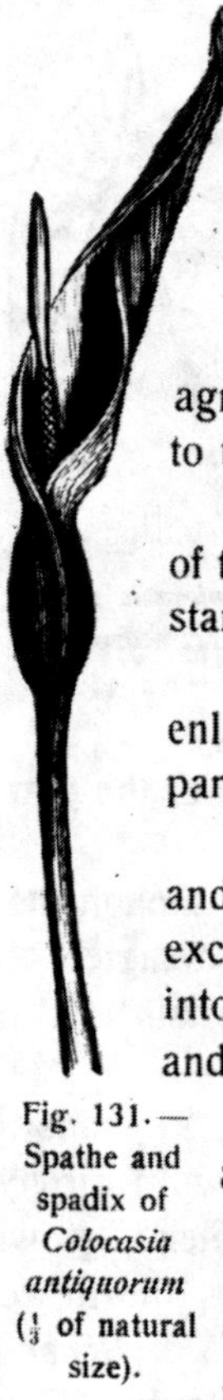

Fig. 131.—Spathe and spadix of *Colocasia antiquorum* (⅓ of natural size).

Fig. 132 —Spathe removed showing spadix with pistillate and staminate flowers.

Besides this appendage and the large, yellow spathe surrounding it, insects are also enticed by:

(a) a *strong smell* of the spadix, which is disagreeable to us, but does not seem to be so to the midges that visit the flowers;

(b) the *nectar,* secreted by the stigmas of the pistils, and the *copious pollen* of the stamens, constituting food for them, and

(c) the *high temperature* in the globular enlargement of the spathe at its lower part, causing them to seek refuge there.

3. After flowering, the spathe fades and nothing remains of the whole spadix, except the lower part, where the pistils ripen into a cluster of **Berries** which are eaten and dispersed by birds.

4. After this the whole plant withers and perishes, excepting the **Tubers,** in which plenty of food is stored up for the next season. When the rains begin again, these plants are among the first to cover the ground with their fresh green. It is also through these tubers that the plant is propagated. Like the Potato, it throws out long underground shoots, the runners or stolons, portions of them being filled with starch and swelled up to form fresh tubers, which eventually grow into separate plants.

Other Aroideæ.

Another well-known species of this order is **Alocasia** *macrorhiza* (*Kan.* Marasaṇige), the gigantic tubers of which are eaten, after the acrid and poisonous juice, characteristic of the family, is driven off by the process of cooking. **Pothos** *scandens* (*Kan.* Aḍike baḷḷi) a climber with small lanceolate leaves and winged petioles; **Rhaphidophora** *pertusa* (*Kan.* Kandoḍi baḷḷi) with large holes in its leaves; **Arisæma** *tortuosum* (the Snake-Lily), with the appendage of its spadix curved outward and then back over the spathe. **Pistia** *stratiotes*, a small herb floating on still waters having rosettes of wedge-shaped leaves.

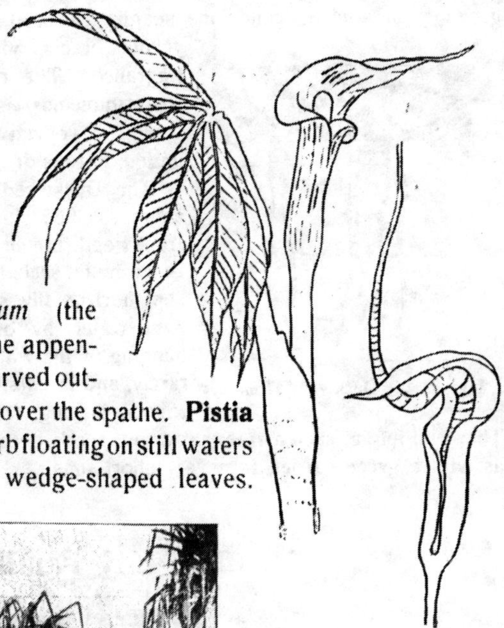

Fig. 133. — Spadix of *Arisæma tortuosum*.

Fig. 134. — Root-end of *Pandanus*.

Allied Families.

The **Screw-Pine** (*Pandanus tectorius*; *Kan.* Kēdage; *Mal.* Kētaki; *San.* Kētaki) belongs to the *Pandanaceæ*. It is often planted for fences on account of its spinous leaves, which are also used for matting. The spines are on the margin, as well as on the underside of the midrib, and the leaves are arranged in three spirals up the stem. Like the Mangrove tree, it forms numerous adventitious roots from the lower part of its trunk, which look like artificial props. At the end of these roots, root-caps can be seen very distinctly. The male flowers, growing on a long, pendulous spadix, enclosed within large, leaf-like, yellow bracts, yield a most delightful fragrance. The fruit borne on separate trees (diœcious) is, similar to the pine-apple, a collective fruit consisting of united, fibrous drupes.

Fig. 135.— Duckweed *(Lemna)*.

The Duckweed Family *(Lemnaceæ)* is another allied family. The **Common Duckweed** (*Lemna globosa; Kan.* Nirāṭa) is a minute, scale-like, green water-plant, kept horizontally on the surface of stagnant water by one long, vertical root hanging in the water. The plant flowers rarely, and is chiefly propagated by side-shoots issuing from the mother plant. They multiply at such a rate that whole ponds become covered with them, as with a green carpet, in a very short time.

34. The Lily Family

(Liliaceæ).

Perennial, usually bulbous herbs with parallel-veined leaves. Flowers radial. Perianth regular, six-leaved in two whorls. Stamens six in two whorls. Ovary superior, formed of three carpels. Fruit a three-celled capsule.

The Glory Lily *(Gloriosa superba)*.

(*Kan.* Karaḍi-kaṇṇu, Šivašakti-baḷḷi. *Mal.* Mēttōnni. *Tam.* Kāndaḷ. *San.* Amrata, Haripriya.)

1. The Gloriosa is one of our most beautiful ornamental plants. "During the rains you find it shooting in the lanes, bordered thickly by huge Euphorbia and Aloe, or in bamboo-thickets. The grace of its form, amidst the stout and ugly plants with their fierce thorns and spikes, and the gaiety and warmth of its flowers amidst the sullen and cold gray of the surroundings have, no doubt, given cause to the superb name the flower bears."

2. It is a fragile, weak **Climber** that sprouts up from a tuberous root-stock. The round, green stem is very slender and long, so that it is obliged to seek the support of other plants or things. For this purpose, it uses its leaves, which are tapering tendrilwise, asking for something to curl round it and climb. And thus the plant raises its top to the free air and full sunlight "unfurling its fire-flowers like banners of triumph".

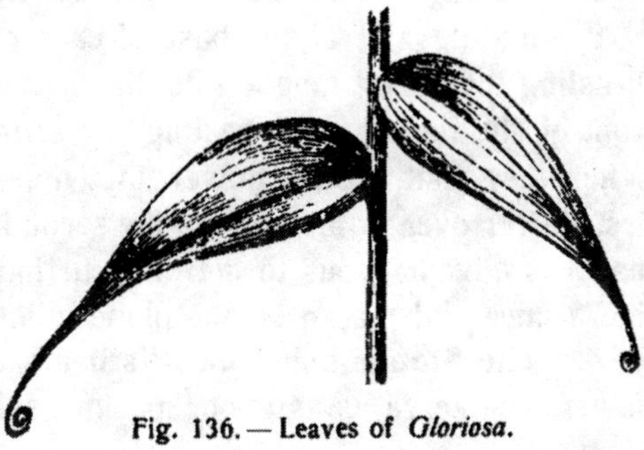

Fig. 136. — Leaves of *Gloriosa*.

3. The **Flowers** are exceptionally beautiful. They are placed in large racemes and hence become visible from a great distance. The floral cover is a perianth of six leaves arranged in two series of three each, the members of each circle alternating with those of the other. This can be easily seen in the green bud which is drooping. When the blossom opens, the perianth-leaves curl back, the six stamens stand out at right angles from the floral axis, and the long, slender style bends away from the hanging ovary at an acute angle and lies in the same plane as the stamens.

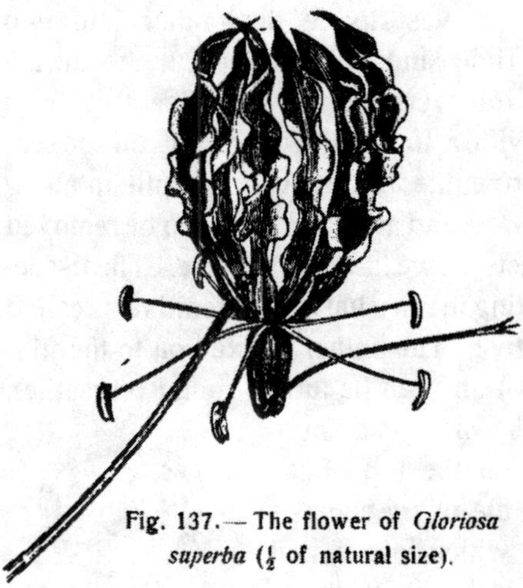

Fig. 137. — The flower of *Gloriosa superba* (½ of natural size).

The flower has no scent, but by the showy colours of its petals it is able to attract insects. The petals are bright-yellow with scarlet tips at first; as they grow older and older, they become darker and darker crimson and bend more and more back. Visiting insects find honey in a hollow longitudinal furrow in the middle and at the base of each petal. They suck it by thrusting their long tongue into the nectary, while hovering in front of the flower, and beating the stamens with their wings. When they visit another flower, they thus carry the pollen of the first flower over to the style of the second. We now understand that it is advantageous to the plant to have its style thus bent like a knee and placed in one plane with the stamens.

4. The **Stamens** and the **Pistil** are unusually large in this flower, and serve us as good specimens for the study of these organs.

It may first be noticed that the stamens also are arranged in two whorls, alternating with one another and with the petals. Examine some stamens in one of the green buds. We find that they consist each of two parts, a filament and a head. The head, called the anther, is grooved both along the face and the back. These grooves divide the anther into two lobes, right and left. If we cut the unripe anther transversely, we shall see two bags filled with a fine yellow powder, the pollen. When ripe, the two anther-bags split up along their edges, and allow the pollen to be removed by insects, as we have seen above. The tissue, connecting the two halves of the anther is called connective. The anther is fixed on to the filament in its back (dorsifixed) and can be turned round (versatile).

Fig. 138. — Diagram of ovary of *Gloriosa*.

The essential part of the pistil is a knob-like vessel, called ovary, on the top of the flower-stalk. In Gloriosa the ovary consists of three leaves or carpels, which fold in and unite at their edges in the axis, so that there are three hollow cells to hold the ovules, which can be seen when the ovary is cut transversely. Even these three carpellary leaves alternate with the three stamens of the inner whorl. So we find that

Fig. 139. — Diagram of Lily flower.

the whole flower consists of five whorls of three leaves each, which always alternate with those of the neighbouring whorl. The ovules are attached to the seams of the carpels, and appear in the transverse section (fig. 138) as axillary growths (axile placentation). The tips of the three carpels unite together, and taper into a slender style which again betrays its threefold nature by a three-parted stigma.

Fig. 140. — The White Lily (*Lilium candidum*).

Other Lilies.

The family of the Lilies to which the Gloriosa belongs, is much celebrated in poetry. The **White Lily** (*Lilium candidum*, fig. 140), growing in temperate climates, is the emblem of purity.

Other plants belonging to this family are the **Onion** (*Allium cepa; Kan.* Nīruḷḷi; *Mal.* Īruḷḷi; *Tam.* Īruḷḷi; *Tel.* Nīruḷḷi; *Hin.* Pyāj); the **Garlic** (*Allium sativum; Kan.* Beḷḷuḷḷi; *Mal., Tam., Tel.* Veḷḷuḷḷi; *Hin.* Lasun); the **Leek** (*Allium ampeloperasum*); the **Dragon Tree,** *Cordyline terminalis,* with copper-coloured leaves, crowded together at the top of a thin cylindrical stem, and panicles of small purplish flowers; *Smilax macrophylla*, a prickly climber with two tendrils at the base of its leaf-stalks, and the **Wild Asparagus** (*Asparagus sarmentosus; Kan.* Halavumakkaḷatāyi; *Mal.* Čadāvēḷikiḷaṅu; *Tam.* Čāttiravēri; *Tel.* Čallagaḍḍalu). The latter has a bundle of many long tuberous roots. Its delicate climbing stem has thorns turned downwards (necessary for climbing, compare Rose, page 45) and its leaves are reduced to minute scales, from the axils of which green leaf-like organs grow, which are homologous with branches and have the function of leaves. The white flowers are small but numerous, and form racemes from the axils of the leaves. The fruit is a red berry.

The Bulb of the Onion deserves our special notice. It is as little a root in the botanical sense of the word as is the tuber of the potato. Cut it through and you will find a flat solid mass at its lower end, from which fibrous roots and a number of concentric, overlapping, succulent scales and leaves (often with a flower-bud in the centre) are developed. This solid substance evidently is the stem of the plant, and the whole bulb is just a bud. The soil in which the bulb is hidden prevents loss of moisture by transpiration, and so its life is preserved through the hot season, though the upper parts, leaves and flowers, all wither and die. The outer scales of the bulb are dry (scarious) and form an additional protective coat over the inner juicy leaves.

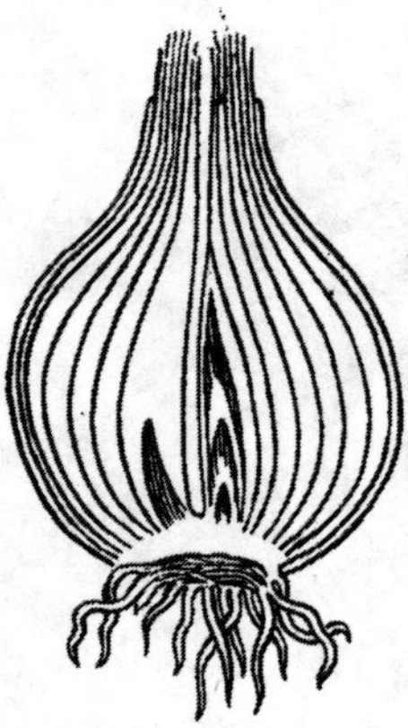

Fig. 141.—Longitudinal section of the bulb of an Onion.

How the bulb represents a food-store of the Onion for its growth in the coming growing season and is thus enabled to take an early advantage of the favourable season, is explained on pp. 91, 92 in the description of the Potato plant. And this is the reason why many of the bulb or tuber-born plants, *e. g.*, the Thunder-lily *(Zephyranthes)*, Crinum Urginea, Iphigenia, Curcuma, and Curculigo, are among the first flowers in the new season.

35. The Amaryllis Family

(Amaryllidaceæ).

Herbs very much like the Lilies. Ovary inferior.

The Asiatic Crinum *(Crinum asiaticum)*.

(*Kan.* Vishamuṅguli. *Mal.* Veḷuttapoḷatāḷi. *Tam.* Tuḍaivāči. *Tel.* Kesaričeṭṭu. *San.* Vishamaṇḍala.)

This is a perennial and evergreen herb with a large elongated bulb, pretty common everywhere, and very conspicuous by its large, glossy, radical leaves, from among which a leafless stalk (scape) two feet long, arises bearing an umbel of numerous white flowers. The flowers are enclosed in bud by a thin bract—the spathe—which breaks up in a slit.

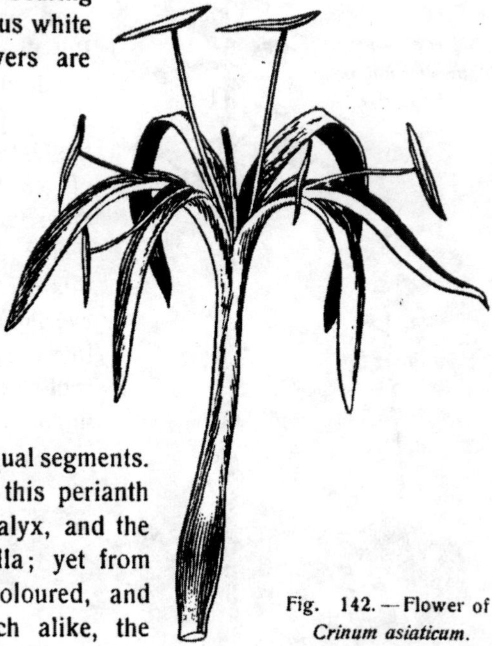

Fig. 142.—Flower of *Crinum asiaticum*.

The perianth consists of two whorls of three petals each, which are all combined into a long tube, spreading at the upper end into six equal segments. The outer whorl of this perianth may be called the calyx, and the inner one the corolla; yet from their being both coloured, and otherwise very much alike, the whole is called the perianth (see also Gloriosa). The six stamens adhere at their bases to the perianth-lobes, and are also arranged in alternating whorls. The ovary is three-celled as in the Lilies, but inferior. The fruit is a berry with one or two seeds.

The bulb and the leaves are poisonous, and are used as emetics.

Other Amaryllids.

These plants bear a close resemblance to the preceding order and are often mistaken for true Lilies. The principal mark by which they are distinguished from the Lilies is the *inferior ovary*. One of their most useful members is the **American Aloe** (*Agave americana; Kan*. Ānekattāḷi; *Mal.* Erōppakaita; *Tam*. Ānekattāḷai). It is a well-known shrub used for fences on

Fig. 143. — The American Aloe *(Agave americana)*. 1/30 of nat. size.

account of its thorny leaves. Its growth is very slow. It requires about ten to fifteen years until it is fully grown. But then a high scape with hundreds of pale, yellow flowers grows suddenly out of the leaf-rosette, to the height of ten to twenty-five feet. After the seeds are ripe, the whole plant dies off. Ere this event takes place, however, numerous shoots issue, like offspring, all round the root-stock. The Agave was imported into India from America where it inhabits the dry deserts of the tropical and subtropical zone. By its succulent leaves, which are covered by a leathery epidermis, it is well suited for such climates (compare Cactus, page 57).—A very strong tough fibre is obtained from the leaves.

Other wild Amaryllids are **Crinum** *ensifolium* (*Kan.* Boggi-kaṇḍa), common on swampy river banks throughout India, and **Curculigo** *orchioides* (*Kan.* Nelatāḷe), a little plant, with about three short leaves from a vertical root-stock and yellow star-like flowers, appearing in June.

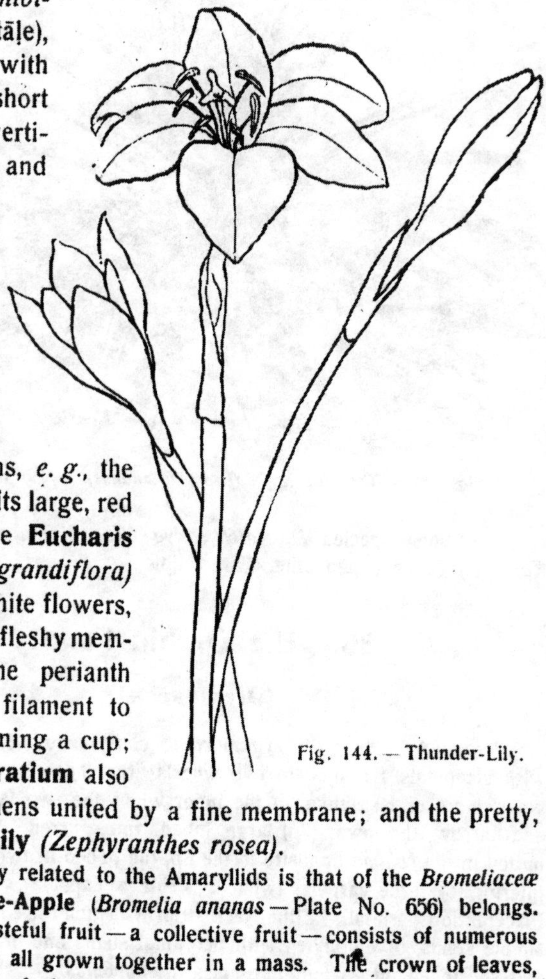

Fig. 144. — Thunder-Lily.

Some of the Amaryllids have very handsome flowers and are prized in gardens, *e. g.*, the **Amaryllis** with its large, red flower-bells; the **Eucharis Lily** *(Eucharis grandiflora)* with fragrant, white flowers, embellished by a fleshy membrane within the perianth stretching from filament to filament and forming a cup; the white **Pancratium** also with its six stamens united by a fine membrane; and the pretty, rose **Thunder-Lily** *(Zephyranthes rosea)*.

A family closely related to the Amaryllids is that of the *Bromeliaceæ* to which the **Pine-Apple** (*Bromelia ananas* — Plate No. 656) belongs. The sweet and tasteful fruit — a collective fruit — consists of numerous flowers and bracts, all grown together in a mass. The crown of leaves, which looks so out of place, growing apparently out of the fruit, belongs really to the flowerless top of the spike, and is capable of developing into a fresh plant. — The fibre of the leaves is used, and the plant is often grown for fences.

Another allied family is the **Yam Family** *(Dioscoreaceæ)*. They have stems twining from left to right, and cordate or digitate, palmately net-veined leaves, in whose axils often brown tubers are borne. The

Fig. 145. — The Pine-Apple *(Bromelia ananas)*. $\frac{1}{12}$ of natural size.

tubers of some species (*Dioscorea sativa; Kan.* Kuṇṭageṇasu; *D. alata; Kan.* Tūnageṇasu; and others) are edible.

36. The Orchid Family

(Orchidaceæ).

Perennial herbs with zygomorphic conspicuous flowers. Perianth with six petals arranged in two whorls; the three sepals of the outer whorl more or less alike, in the inner whorl the two lateral petals alike and narrow, the lower (lip) large, lobed, and spurred. Stamen and style united into a *column* opposite to the lip, the pollen in two masses. Ovary inferior, of three carpels, twisted. Fruit a capsule with many seeds. Placentation parietal. — Either terrestrial with tuberous roots and annual, simple stems, or epiphytes with perennial stems and branches, thickened and shortened into a bulb-like mass *(pseudobulbs)*.

The Round-leaved Habenaria *(Habenaria rotundifolia)*.

(*Kan.* Nelatāvare, Oreletāvare.)

1. When the rains begin to moisten the ground after the long drought from October to May in Western India, the

terrestrial Orchids, which are to be found in hills and dales, also awaken and send up their shoots to the light and air. For, their **tuberous roots** enable them to form leaves and stems at once from the reserve food contained in them. This food is mainly starch, and is so rich that, in some kinds, it can be used for the preparation of food for man, the "salep".

If the plant is dug out early in the season, a young bud can be seen in the axil of one of the dry scales that surround the shoot. This bud gradually swells and becomes a tuber, like the old one, so that, if we examine the plant at about the time of flowering, in July, its size is equal to that of the old one, the mother-tuber, which then already shows signs of shrivelling. If the plant is examined once more when the fruit is ripe, the mother-tuber will be found brown and withered, whereas the daughter-tuber will be firm, light-coloured, stuffed out, and provided with a bud at its upper end. These phenomena are the same as those which we have noticed in the Potato tuber (page 91) and the bulb (page 154). The reserve food in the tuber is used for the building up of the leaves, flowers, and fruit of the plant, during which process the plant stores fresh food in a new tuber for use in the following year, and the old tuber being exhausted falls off and decays.

Fig. 146.— Tubers of *Habenaria rotundifolia* at the time of fruiting (½ of natural size).

2. The sprout of *Habenaria rotundifolia* forms one or two round **Leaves** (fig. 146), pressed close to the ground. Their resemblance to the leaves of the Lotus-plant has given origin to the vernacular name *Nelatāvare,* which means Ground-lotus. They are quite *smooth*. Being surrounded by the moist air of the monsoon, and growing in swampy soil which affords an ample supply of water, the plant can dispense with the protective coat of hairs which we frequently find on the stems and leaves of

plants growing on dry soil and during the dry season. The leaves are *dark-green*, which is likewise referable to the habitat of the plant. Dark-coloured things, we know, absorb more heat than light-coloured ones. The higher the temperature of the leaves of a plant rises, the more abundant is the evaporation and hence the growth (see II. Part, Transpiration). The damp surroundings, which naturally check the evaporation, are thus compensated for by the dark colour of the leaves.

Fig. 147.— Flower spike of *Habenaria rotundifolia* (2 times magnified).

The thick, fibrous roots which spring from the stem immediately over the tubers, are few and small; they are sufficient for the supply of water from the soil which is constantly humid during the time of its growth.

3. A slender, leafless stem, called scape, about six inches high, rises from the radical leaves and lifts the few **Flowers** over the tips of the grass around it. They thus become visible to passing insects which have to fertilize them.

Each flower is attached to what seems to be a short twisted flower-stalk rising from the scape in the axil of a conspicuous bract, but it is really the ovary, which can be proved by cutting it through and thus laying the ovules inside bare.

The perianth is composed of two sets of three leaves each. The middle petal of the outer set and the two upper or lateral petals of the inner set are bent together and form a hood to protect the inner organs. The two other outer petals are expanded, and the middle petal of the inner set, generally called the lip, is divided into three narrow lobes and drawn out into a long, slender spur. Close to the entrance to the spur we find the stigma of the pistil with a sticky surface, and above it, the sessile stamen which contains two pollen-masses in a pouch each; these look like two small clubs, and end in a gummy disk below, covered under a small, knob-like projection.

This very peculiar structure of the Orchid-flower is fully understood only when we study the mode of its *pollination by bees.*

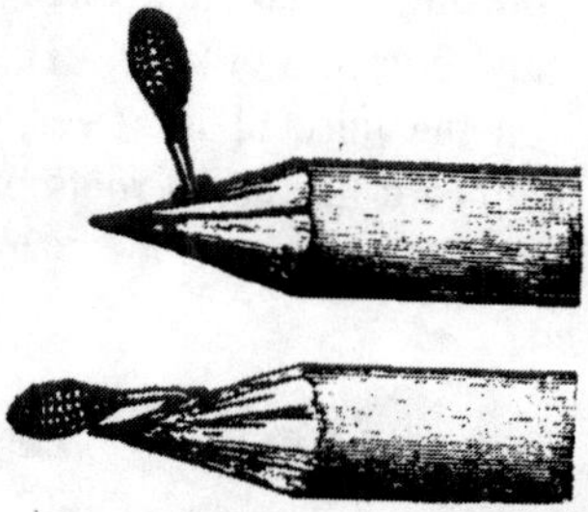

1. 2.

Fig. 148.—Pollen-masses at the end of a pencil: 1. immediately on the withdrawal of it, 2. one or two minutes afterwards.

The flowers are small, indeed, but their white colour makes them visible as they are raised by their stems above their surroundings. A bee seeing them alights on the expanded under-lip which affords a convenient landing place. It then stretches its proboscis into the spur in search of nectar. Just at the entrance to the spur are the projections which contain the sticky disks of the pollen-clubs under small lids. As the insect touches these with its head, they break up, and at once the sticky disks settle on the forehead of the insect. On leaving the flower, the pollen-masses are drawn out of their pouches, and the bee flies away with them. If this process is imitated by gently inserting a pointed pencil into the spur and withdrawing it, one can see that the clubs which are erect in the beginning bend forwards after a minute or two (fig. 148). The same happens, of course, when the pollen-clubs are on the head of the bee. When it thus visits another flower, the pollen-masses must touch the stigma of it, and this will detach some or all of the pollen from the bee's head. The flower is thus fertilized.

4. The **Fruits** are capsules containing numerous powdery seeds which are shaken out and dispersed by the wind when the capsules split into six valves.

Other Orchids.

The Orchid Family comprises more than five thousand species. Many of them have very showy flowers. They flourish largely in the hill tracts of India, and as they do not get sufficient light on the ground in a forest, they frequently settle on the trunks and branches of trees. They are, however, not parasitic, preying on the juice of the trees, like the Loranthus, but simply throw out cord-like aërial roots by which they attach themselves to the

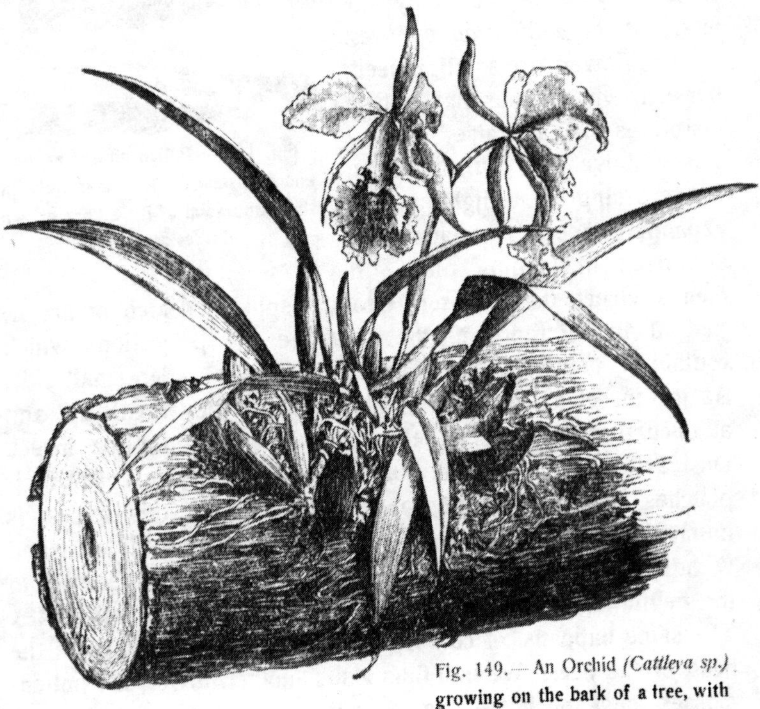

Fig. 149.—An Orchid *(Cattleya sp.)* growing on the bark of a tree, with pseudo-bulbs (½ of natural size).

bark. They feed on the dust blown by the wind into the fissures of the bark, and on the rain and dew falling on them. Such a habitat must be unfavourable, indeed, during the rainless season; but in many genera of the Orchids this is compensated for by the lower joints of the stems or leaves becoming much thickened and fleshy (pseudo-bulbs), so as to resemble bulbs, where they store up each drop of water they can get. (See also page 131.)

A very common type of these epiphytic herbs is **Vanda** *Roxburghii* (*Kan.* Marabāḷe), frequent on Mango and Banyan trees. It has axillary racemes of sweet-scented, chequered, yellowish and purple flowers and succulent, recurved leaves in which water is stored up.

Dendrobium *barbatulum*, another epiphyte and flowering in December, has racemes of little cream-coloured flowers, and stores up water and food in its grey succulent stems which are leafless during the greater part of the year.

Vanilla *planifolia*, a climber, is cultivated for its fruit, which is taken when unripe and yields the well-known aromatic essence "Vanilla".

The pretty snow-drop like Orchid commonly found on the trees of the Ghauts is *Cœlogyne flaccida*.

37. The Ginger and Arrowroot Family

(Zingiberaceæ).

Perennial herbs with radicle leaves, which are pinnately nerved from the mid-rib. Flowers zygomorphic. Sepals three, free or united. Petals three, tubular. Stamen one with five petaloid staminodes. Ovary inferior, three-celled. Fruit a capsule.

The Ginger Plant *(Zingiber officinale).*

(*Kan.* Çuṇṭhi. *Mal.* Iṅči. *Tam.* Iṅji. *Tel.* Çuṇṭhi. *Hin.* Sōṇṭ.)

1. This is a herb with a creeping **Root-stock** *(rhizome)* containing a great deal of starch and an ethereal oil, which gives the tubers an agreeable aroma and a warm, bitterish taste. It is cultivated, and the root-stocks are used as spices and in medicine.

If we examine the root-stock of a Ginger plant, as it is sold in the bazaar, we shall see that it contains buds or "eyes" at its ends and some scars in its middle portion (fig. 150). If it is planted, the buds will produce leaves and, perhaps, flowers, whilst new buds will be formed at the further end of the root-stock. This *underground stem* thus creeps along under the soil producing fresh buds every year, slowly moving away from the spot where the former plant grew and thus always seeking fresh, unused soil. The old portion marked with the scars of withered plants decays, and as the side-buds similarly creep along in opposite directions, independent plants are produced eventually. The branching of the root-stock is, therefore, *a means of vegetative propagation,* and this is by no means an unimportant one as the Ginger plant rarely flowers to produce seed.

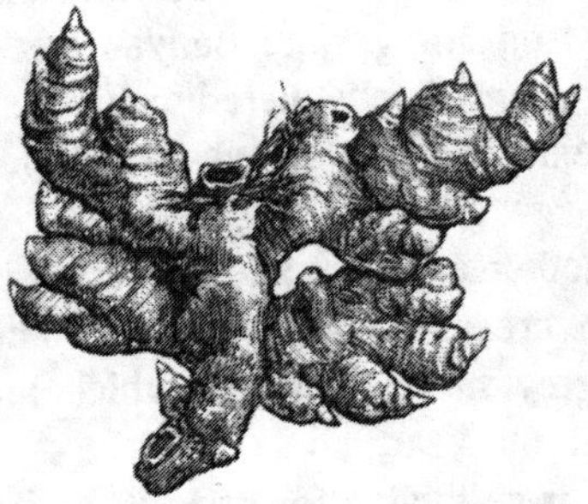

Fig. 150.—Rhizome of Ginger *(Zingiber officinale).* ½ of natural size.

2. **Leaves.**—The Ginger plant grows during the monsoon and so does not require a protective coat of hairs on its stem and leaves (*cf.* Habenaria, page 159), or other means of checking the process of evaporation, such as a limited surface of the leaf-blades. It can, therefore, develop large and long leaves. The leaves, like all the other parts of the plant, contain the volatile oil, which we have noticed in the root-stock, that gives them a fine aroma when bruised.

3. The **Flowers** appear not on the leafy stem, but are produced on separate scapes that rise from the root-stock, a little removed from the leafy stem. They form a spike at the top of the scape, being supported and protected by imbricated, concave bracts.

The perianth of the flower is tubular and has a double border of three lobes each. Within the perianth there are six "leaves" in two sets, belonging to the staminal series of which, however, only one bears an anther. Of the rest two are reduced to minute teeth, nestling around the base of the style, whereas one is enlarged to a violet, petal-like lip, and two are found as short teeth or lobes at each side of the lip. The anther-bearing stamen is drawn out into a purple sheath clasping the upper part of the style, whose funnel-shaped stigma overtops all parts of the flower. The arrangement of these inner organs clearly shows that self-pollination is impossible.

Other Plants of the Ginger and Arrowroot Family.

Many plants of this and the allied family *Marantaceæ* are useful as food stuffs, spices, and dyes. Arrowroot flour is made from the root-stocks of **Maranta** *arundinacea* and of **Curcuma** *angustifolia* (*Kan.* Kūve; *Tam.* Kūkai). The seeds of **Cardamom** (*Elettaria Cardamomum; Kan.* Ēlakki; *Mal.* Ēlam; *Hin.* Ēlāči) are used as spices. The root-stock of **Turmeric** (*Curcuma longa; Kan.* Arisina; *Mal.* Maṅṅaḷ; *Tam.* Maṅčal) yields a common dye.

Fig. 151. – Flower of *Costus speciosus.*

Garden plants are **Costus,** *Costus speciosus* (*Kan.* Pushkaramūla; *Mal.* Čannakkūva; *Tam.* Kōshṭam), a most elegant-looking plant, remarkable for its spirally ascending stem and the two white glossy petal-like stamens, beside which the three small red sepals and the three white petals proper sink into insignificance; and the **Indian Shot** (*Canna indica; Kan.* Kēḷahū; *Mal.* Kāṭuvāra; *Tam.* Pūvāḷai).

38. The Banana Family

(Musaceæ).

Herbs very much like the Zingiberaceæ, differing chiefly in the arrangement of the petals and stamens.

The Banana or Plantain *(Musa paradisiaca).*

(Plate No. 636.)

(*Kan.* Bāḷe. *Mal.* Vāḷa. *Tam.* Vāḷai. *Tel.* Araṭi. *Hin.* Kēli. *San.* Kadali.)

1. **Stem and Leaves.**—The stem, formed from the sheaths of the leaves, tightly rolled round each other, attains a height of

from fifteen to twenty feet. The leaves spring from a short tuberous root-stock, which buds out at its sides. As a matter of fact, the Banana plant is just a much elongated bulb, the scales and leaves forming not a globose body as in the onion, but a long cylinder (*cf.* Bulb, page 154).

The leaves are *very large,* six to eight feet long and two feet broad. In bud they are rolled up from one side. They would offer much resistance to the wind, the pressure of which the weak stem could hardly withstand. To prevent the plant being thus overturned, nature corrects herself in a very simple manner: the leaves have a strong, fleshy mid-rib, from which the veins run to the margin at right angles, and they *split readily when swayed by the wind;* the leaf now acts like a pinnate leaf, the various parts letting the wind pass between them, and thus lessening the resistance (*cf.* Cocoanut tree, page 139). The mid-rib and the stem-clasping sheath contain numerous large cavities, the result of the very quick growth of the plant, there being hardly time enough to provide the food stuff required to form so much solid tissue.

2. **Flowers.**—The flowers are borne on a long scape running through the leaf-sheaths as a white cylinder. It inclines downward by its own weight in a graceful curve. The flowers are arranged in whorls or clusters, and each whorl of flowers is protected under an ovate, concave, leathery bract (spathe), crimson on the inside and with a pale bloom on the outside. Eight or more of these, nearest the base of the huge, drooping panicle, embrace a double row of ten to sixteen flowers which are fertile. With the maturity of each successive row of flowers, the spathe reclines and falls off, and the fruit appears. The rest of the whorls—and they are very numerous—expand in succession for two or three months, and contain similar double rows of flowers which, however, do not bear fruit, but fall with their spathes.

The perianth consists originally of six petals arranged in two whorls (compare Lilies Amaryllids, Orchids and Zingiberaceæ). The three outer sepals and two petals of the inner whorl are united into a tube with a slit throughout its length, in which the

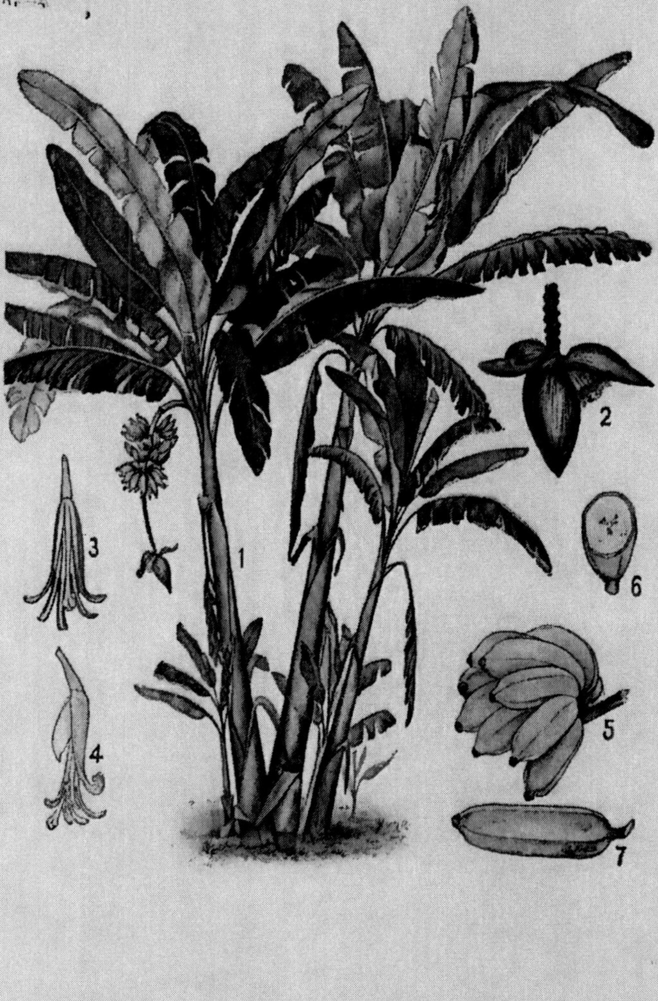

Fig. 152.—BANANA OR PLANTAIN *(Musa paradisiaca)*.

2. Head of spike. 3. Single flower (front petal removed). 4. The same, petal not removed.
5. 6. 7. Fruit.

third petal of the inner whorl is seated. The five teeth of the floral tube betray its composition. The stamens are ordinarily five in number, the sixth being abortive. Plenty of nectar is secreted from the base of the flower, and bees are often seen swarming about the flowers to fetch the nectar and, in return, to fertilize the flowers. The stamens of the flowers in the eight or more first whorls are sterile, whereas their large pistils, crowned with a clammy stigma, are fertile. Inversely, the pistil is sterile and the stamens are fertile in the flowers at the tip of the spike, which, therefore, are dropped after flowering.

3. The well-known **Fruit** is an oblong berry, composed of three carpels tapering at each end, and of a fleshy consistency (fig. 152, 6). The numerous seeds in it are usually not developed. The plant is, therefore, not propagated by seeds.

With the production of fruit the growth of the plant ceases. The life-time is from nine months to three years, and under good conditions ordinarily about a year. During this short time the Plantain develops into that stately and magnificent plant, a phenomenon which is unique even in the tropics. The plant is, therefore, highly esteemed by the Hindus as the emblem of plenty and fertility, and is, as such, in constant requisition at their marriages and other festivals for ornamenting the entrance of houses and temples.

Besides the fruit, which is eaten in many ways, and also dried and made into flour, the fibres of the sheaths and of the leaf-stalks are used. The leaves are used by Brahmans instead of plates.

4. In the life-story of this plant, which has been cultivated for ages, we see a great abnormality. *The seeds of the plant are not properly developed.* This is the result of man's interference with the natural growth of the plant, as can be observed in the wild species, *viz., M. sapientum,* or in *Musa superba*, a plant often grown in gardens for its gigantic, ornamental leaves. The fruit of this kind bears numerous black seeds embedded in very little pulp, which produce healthy plants. By constantly preferring and selecting sorts with richer pulp and paying no attention to the seeds which were useless to man and not necessary for reproduction, a variety was eventually obtained which produced

nothing but pulp in the fruit. (Compare the inferiority of the seeds of the Potato plant, page 93.)

39. The Grass Family

(Gramineæ).

Herbs (except the Bamboo), with a jointed, hollow, leafy stem, called *culm*. Leaves entire, straight-veined, sheathed. Flowers glumaceous, *i. e.*, consisting of dry and scaly *glumes*. Stamens generally three (six in Rice). Ovary crowned with two feathery styles. Fruit one-seeded, indehiscent, dry. Seeds endospermous, mealy and often nutritious.

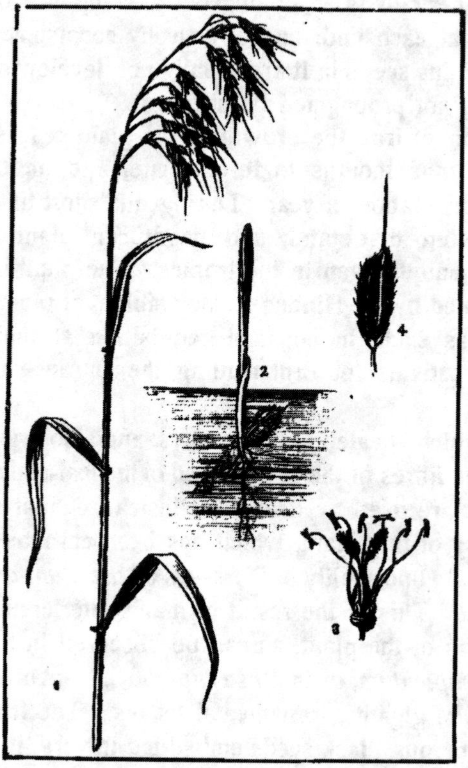

Fig. 153.— The Rice plant *(Oryza sativa)*.
1. Flowering plant. 2. Seedling. 3. Flower. 4. Grain.

The Rice plant
(Oryza sativa).

(Plate No. 626.)

(*Kan*. Bhatta. *Mal*. Ari. *San*. Vrihi.)

1. Importance.— Rice is the principal food of about one-third of the population of the world, and is, therefore, the most useful and most important of all cereals. It is found wild in some parts of India, but has from time immemorial been cultivated throughout the warmer regions of the world. What bread is to the people of the temperate zone, boiled rice is to those of warmer countries.

2. **The Grain and its Germination.**— The rice grain is a small, grayish-yellow thing enclosed in two hairy husks or glumes, the larger of which is five-nerved and sometimes terminates in a bristle (awn). The grain itself must not be regarded as a seed corresponding to that of the Gram or Cucumber: it is a fruit consisting of fruit-cover (pericarp) and seed, like the achene of the Sunflower; but the pericarp is here adnate to the seed.

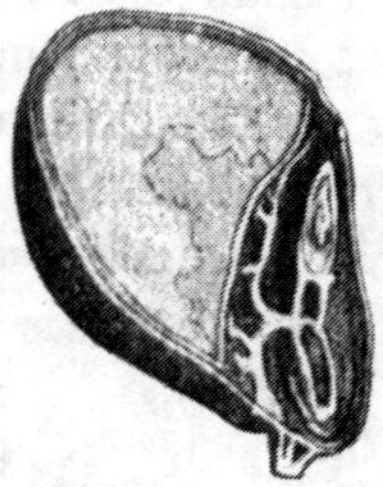

Fig. 154. — Longitudinal section through a grain of Maize.

In the seed of the Gram (page 30) we have found the plumule lying between two thick seed-leaves or cotyledons. The structure of the grain of Rice is quite different. Here we have a very small germ at one end of the grain, the remaining part of it being filled up with a mealy substance, called *endosperm*. The embryo at the lower end and on the side of the grain does not consist of two cotyledons with the plumule lying between them, but of a small oval body enclosing plumule and radicle. Soak a few grains and cut one of them lengthwise. You will then, with the help of a hand-lens, be able to see the parts of the embryo, and as the grains of Maize are larger, the structure of the embryo may easier be seen in that seed. There is a series

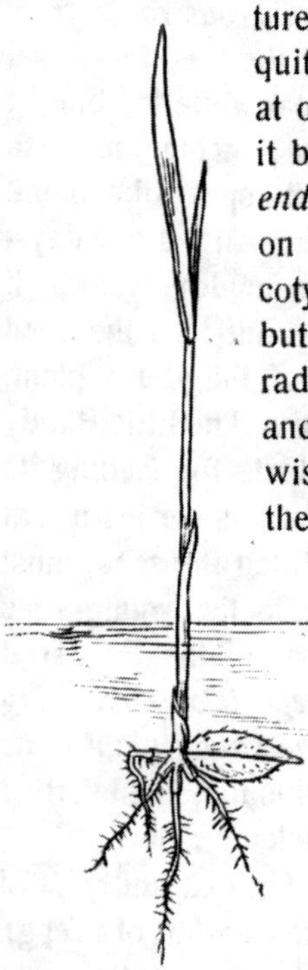

Fig. 155. — Young Paddy plant.

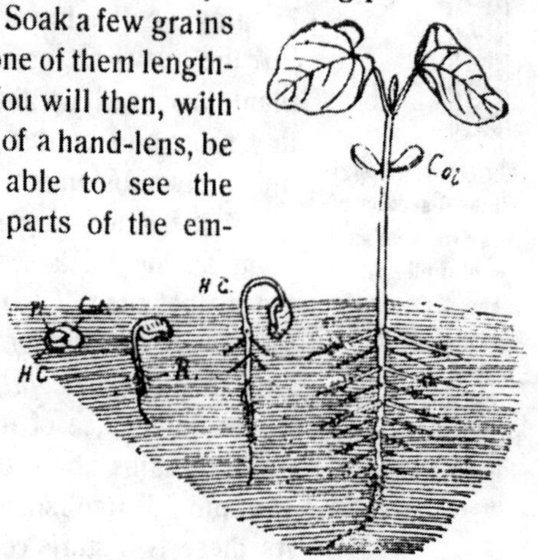

Fig. 156. — Young Gram plant.

of sheathed rudimentary leaves pointing upwards—the plumule,—and another structure pointing downward—the radicle. The remaining part which encloses plumule and radicle and adjoins the endosperm must morphologically be considered as the cotyledon, though in shape and function it is very different from the cotyledons in the Gram. On account of its shield-like shape it has been called scutellum.

If the seed is sown, it will be seen, as in the Gram, that the rootlet of the embryo makes its appearance first. But, whereas the root of the Gram elongates forming a tap-root from which side-roots are thrown off, the radicle of the young Paddy plant ceases growing, and sends forth *numerous roots from small sheaths* at the base of the seed forming a bundle of fibrous roots. Meanwhile the bud of the plantlet grows upwards piercing the earth or mud lying over it, with its spear-like point.

Fig. 157.—Longitudinal section of a Grass-culm. *g. C.* Full grown and *t. C.* tender part of an internode. *N.* Node. *Sh.* Sheath of a leaf. *Sh. Sw.* Swelling of the leaf-sheath above the node.

Rice is sown in wet land. After a few days, thin blades peep out of the muddy water. If we pull up one of them, we still see the seed hanging on the lower end of the young plant. But the husks are now empty. The little Paddy plant has used up the contents for feeding its tiny roots and blades. From this we learn that the mealy substance which filled the seed, must be of the same importance to the young Rice plant as the thick seed-lobes of the Pea seed are to that plant. It is *a deposit of nourishing matter for the support of the young plant in its first stages of growth,* and this food is absorbed by the inner face of the cotyledon.

It contains about 80% of starch and 7% of albuminoides (substances like the white of an egg). As these two stuffs constitute the essential parts of human food, we can understand why rice is so valuable to man.

3. **Stem.**—After some time the little plants, sown closely in small seed-beds, are taken up and planted out again. If this

were not done the plants would choke one another. This process also forces the plant to form more numerous and stronger roots which are helpful for a rich crop. Adventitious roots grow freely out of the lower nodes of the stem.

The plants soon produce stems, called *culms,* usually three to four feet long. Although they are very thin, they are strong enough to bear the weight of their leaves and that of the grain in the panicles. They are *elastic* and, when blown to and fro by the wind, suffer no injury. As in the stems of the Labiatæ (p. 105), it is the outer part of the stem that suffers the greatest pressure when bent. Those plants, therefore, have the four edges of their stems strengthened by strong fibres. In the grass culm *a round tube* is formed by such strong fibres. The tissue in the middle disappears, as it has to bear no pressure when bent: *the culm is hollow.* Only the *nodes are solid* and divide the culm into various parts, called internodes. This serves to strengthen it. That part which is to suffer the greatest tension, namely *the base* of the culm, *has its nodes nearest together* in order to make it stronger at that particular part.—Paddy has its roots under water; and these must be supplied with air (see Lotus, p. 3), which is done by air-chambers and canals running through the tissue of the stem and of the leaf-sheaths.

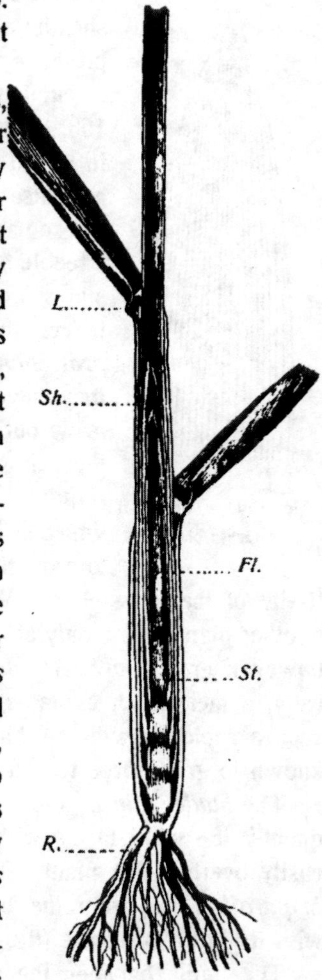

Fig. 158.— Longitudinal section of a young Grass-culm. The stem *(St.)* with its leaves and inflorescence *(Fl.)* grows under the protection of the sheaths *(Sh.)* of the older leaves. *L.* Ligule. (Natural size.)

4. **Leaf.**—Each leaf consists of two quite different parts, the *sheath* and the *blade*. Where both join, there is a small membraneous appendage, called the *ligule*. The sheaths arise from the nodes and form tubes protecting the stem. If a young Paddy plant is cut lengthwise, as in the illustration fig. 158, the sheaths will be seen to form a case, in which the stem *(St.)*, the younger leaves, and also the flower-buds *(Fl.)* are enclosed. These parts are so extremely tender that even a feeble breeze could break or the heat of the sun scorch them. *The sheaths* of the older leaves that rise above them thus *form a protection to these tender parts*. Only when they have attained sufficient strength, they grow out of their protecting cover. Besides, they afford the culm more support, which it requires very much, as those parts of it which lie immediately over the nodes continue to grow for a longer period. This is a peculiarity of the grass-culm. We know that the stems and stalks of other plants grow only at their ends; *the stems of the Grasses*, however, *grow above every node* and thus lengthen out telescopically, a fact which explains why Grass plants generally have such a rapid growth. A bamboo stem, for instance, has been known to grow three feet in twenty-four hours.

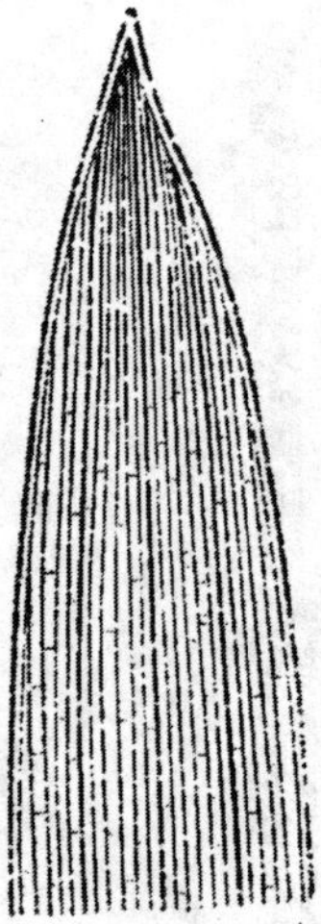

Fig. 159.—The tip of a Grass blade.

The *blade is linear* and waves like a flag in the wind. Consequently the wind meets with no resistance from it and cannot easily overturn the plant. The *nerves* or ribs of the leaf run *in parallel lines* from the base to the tip of it, in accordance with its linear structure (fig. 159).

The ligule between the sheath and the blade prevents the rain-water running down the leaves from entering under the sheath and thus rotting the tender parts of the culm.

If the leaves are drawn swiftly through the hand, they cut. This is due to the presence of *silica*, a substance, which constitutes some of the hardest stones like quartz and flint, and is deposited in large quantities in the cell-walls of the epidermis of the leaf and stem. This is certainly not without its good purpose. The hard silica is not only a means of support to the plant, but protects it also from the attacks of a number of small animals, such as snails, caterpillars and insects, that would feed on the leaves and stalks but for this substance which interferes with their feeding. Bigger animals, like cattle, do not mind it. But there are genera of Grass on the Ghauts which because of their coarseness are disliked even by cattle.

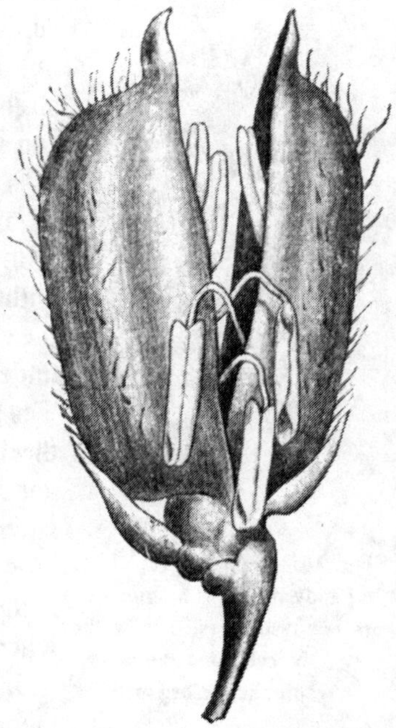

Fig. 160. — The flower of the Paddy plant (very much enlarged).

5. We now come to the **Flower** and **Fruit.** The flowers are supported by a panicle which bows down with the weight of the ripening grains. The short lateral flower-branches are wiry and bear one-flowered, stalked spikelets. Each spikelet consists of two very short scaly bracts at the base (fig. 160), which reasoning from the analogy of other grass flowers represent two abortive flowers, and two larger dry scales (glumes) closing over the naked flower at the top of the spikelet. The outer of these two, the glume proper, is boat-shaped embracing the opposite smaller one (palea) with its incurved edges, and sometimes ending in a short bristle, the awn. Between the glume and the palea are two small glabrous bodies, called lodicules. The inner organs consist of six pendulous stamens and the pistil with two feathery styles (fig. 152, 4).

We know that, if seeds are to be produced, the pollen of the stamens must fall on the style of the pistil. The two glumes that enclose these organs, therefore, open when the organs are sufficiently developed. There is a wise arrangement in the Grass flower in order to make it open in proper time. It would be quite useless for it to open during the rain, for the rain would wash the flower-dust away, and the pistil could not be fertilized. The two lodicules swell when the sun shines warmly on the flowers and press the glumes slightly open, so that the six stamens can protrude their anthers. The wind then takes the pollen and carries it to the next flower, which catches it readily with its *feathery styles*. After this the anthers fall off, the lodicules shrink, and the glumes close again, like two lids, and under their protection the fruit ripens. The husks remain on the seed even when the latter drops from the plant. They keep the seed nice and dry. When it is sown in wet land and has absorbed moisture, the husks keep the moisture within them for the growth of the embryo even when the soil becomes dry for a while.

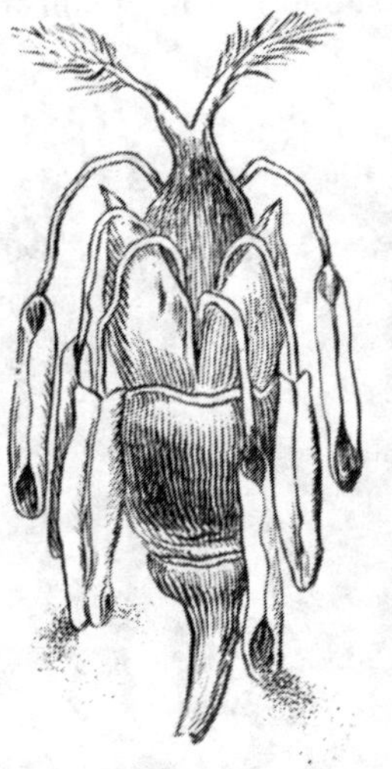

Fig. 161 — The flower of the Paddy plant. The glumes are removed so as to shew the two lodicules and the inner, reproductive organs.

6. **Enemies.**—From the seed-bed to the granary this plant is surrounded by a host of enemies, against whom the cultivator has to wage war: numberless ill-weeds *e. g.*, the much-dreaded *Ischœmum rugosum,* misappropriate its light, space, and food in the field; parasitic Fungi, of which Rust, Smut and Bunt may chiefly be mentioned, settle on culm, blade and flower; grubs and other larvæ of several insects feed on its roots; birds and other animals eat away the ripening grain; and even in the godowns rats, white-ants, the grain weevil, and other uninvited guests can do a great deal of damage to the stored rice.

Other Grasses.

1. **Wheat** (*Triticum vulgare,* Plate No. 626—*Kan.* Gōdi; *Mal.* Kōtampam; *Tam.* Gōdumai; *Tel.* Gōdhumalu). Among the cereals, cultivated in India, Wheat comes next in importance to

Rice. It yields a fine, white flour, which is used for baking bread and for preparing starch.

2. **Maize** or **Indian Corn** *(Zea mays,* Plate No. 631—*Kan.* Mekkejōḷa; *Mal.* Ponticcōḷam; *Tam.* Mokkāičōḷam) is of American origin, but is now largely grown in India. As the few roots developed under the ground are not sufficient to fix the robust culm, with its long, ribbon-like leaves, strong enough in the ground, the lower part of the stem forms adventitious roots, (Plate 625, fig. 1), which, like the ropes of a flag-staff, hold it firmly. Unlike all other grasses, Maize is monœcious. The male flowers (fig. 2, 3, 4, 5) are at the top of the plant in a large panicle, the female (fig. 6, 7) are produced lower down in the axils of the leaves and form dense spikes, enclosed in numerous sheaths which protect the coverless pistillate flowers. But as the stigma must be exposed to the wind (why?), the styles are drawn out into long filaments, which protrude from the top of the sheath like a long, silky tassel. The large, mostly yellow grains are densely packed on a thick core, thus forming what is known as a cob.

Maize is a good food for men and domestic animals. The stalks are valuable as fodder, especially when the cobs are disposed of in the green state.

3. **Other Cereals** grown in India are

Indian Millet (*Sorghum vulgare; Kan.* Biḷē jōḷa; *Mal.* Čōḷam);

Barley (*Hordeum hexastichum; Kan.* Javegōdi; *Tel.* Yavalu; *San.* Yava);

Oats (*Avena sativa; Kan.* Tōkēgōdi);

Ragi (*Eleusine coracana; Kan.* Rāgi);

Little Millet (*Panicum miliaceum; Kan.* Bagaru);

Italian Millet (*P. italicum; Kan.* Navaṇe); and

Panicum *frumentaceum* (*Kan.* Sāme).

4. **Sugar-cane** (*Saccharum officinarum,* Plate No. 631.—*Kan.* Kabbu; *Mal.* Karimbu; *Tam.* Karumbu; *Tel.* Čeruku; *Hin.* Gannā).—This plant is indigenous to India and yields a higher proportion of sugar than any other plant cultivated for sugar. The perennial root-stock produces numerous, solid culms growing to a height of ten feet which bear tufts of leaves and a spreading

panicle of glumaceous flowers (fig. 162, 2 and 3) at their end. As the old leaves fall off, and leave scars, the nodes of the culms become visible. The internodes (fig. 162, 4) are often striped with various shades of red and green.

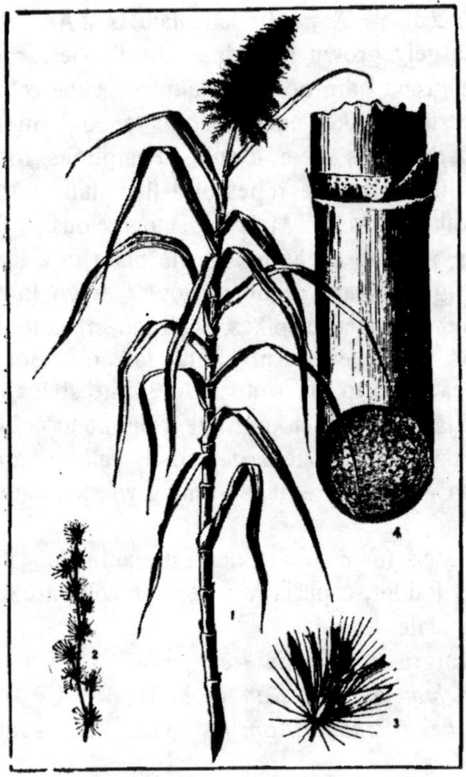

Fig. 162.—The Sugar-cane *(Saccharum officinarum)*
1. Flowering cane. 2. Part of panicle. 3. Single flower.
4. Culm, showing solid consistency and node.

The canes are cut before flowering, as the juice is then in greatest perfection. They are divested of their leaves and of their tops, which contain little or no sugar, and are then crushed in the sugar-mill so as to obtain the sweet juice. This is mixed with lime, boiled down, clarified, and then cooled, and taken to the market as refined sugar.

5. **Bamboo** (*Bambusa arundinacea; Kan.* Biduru; *Mal., Tam.* Mūṅgil; *Tel.* Veduru; *Hin.* Bhasā).—1. The larger type of this genus grows either in *isolated clumps* (fig. 163), or in extensive forests, allowing no other trees to associate with it. The smaller species are found as underwood scattered in the forests.

2. The numerous **Stems** of a Bamboo clump (fig. 163) spring from a large underground stem. In sandy and dry soil they attain a height of about ten feet, but in muddy soil they may

become a hundred feet high. The culms are about three inches in diameter, with the nodes rather near one another.

Fig. 163. — A Bamboo Clump *(Dendrocalamus giganteus)*.

The stems in the centre of the clump are vertical, those at the outer edge hang over giving the whole group a graceful appearance.

The lower part of the stem is commonly branchless, and often spiky from rudimentary hard branchlets. The lowest nodes are set with a ring of numerous adventitious roots, which when they do not reach the earth sometimes turn into spikes. The upper part of the culm bears branches issuing from the nodes in semi-whorls. When these are young, they point upwards and enable the stem to find its way up through the thicket; when old, they spread, bend down, and lean on any stems below them. When the wind moves the culms, they rub against one another and produce a groaning sound which is familiar to people who live near bamboo-thickets.

3. In August and September **Young Stems** appear from the root-stock. The rapidity of their growth is astonishing. The rate of it can be ascertained by taking measurements every day. These tender culms produce, at their nodes, imperfect leaves, *viz.*, sheaths with no blades, which serve as protective covers. When the internodes are full-grown these become useless and drop. The leaves at the ends of the culm are complete. These, too, are shed in the dry season, a peculiar feature in the Bamboo, unlike other grasses.

4. The **Flowers** are produced in branched spikes clustered round the nodes. They are generally monœcious. Stamens six. Style with a feathery stigma. Bamboos flower rarely. But when they flower, they do so throughout the whole district at the same time. It has been observed that thirty-two years lapse from one flowering year to the next; for the years 1804, 1836, 1868 and 1900 have been the years when the bamboo was in flower during the last century in Southern Mahratta, Kanara, Malabar and Coorg. After flowering the root-stock seems to be entirely exhausted, and produces but small and weak culms. Slowly but gradually its vigour is restored so that only after the lapse of thirty-two years flowers and seeds can be produced again.

5. The **Uses** of the Bamboo are nearly as manifold as those of the Cocoanut tree. The stems are used for building houses, sheds, bridges, for manufacturing all sorts of furniture and household articles; young sprouts are eaten as vegetable, and the seeds afford a nutritious food like rice.

6. **Spinifex** (*Spinifex squarrosus; Kan.* Rāvaṇa gaḍḍa).

1. **Habitat.**—Every one who has once strolled along the

sandy seashore of India is well acquainted with the spiny, bluish grass Spinifex and its spherical inflorescences. Spinifex sometimes covers the outer dunes in innumerable, apparently isolated tufts. On closer examination it will be found that they are not single but connected with one another by horizontal runners

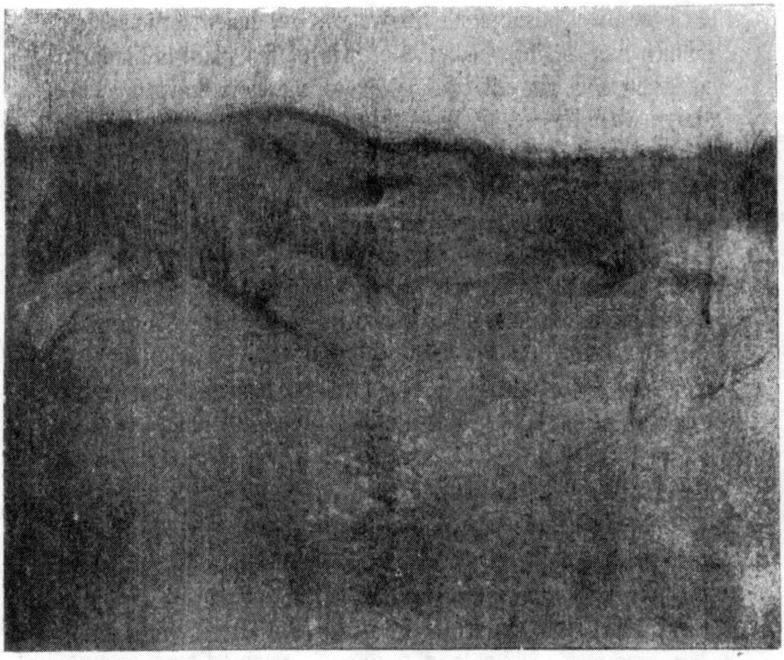

Fig. 164.—Dunes with *Spinifex squarrosus.*

hidden under the sand, as thick as a goose-quill or a finger. These produce roots and side-shoots at their nodes. The leaves owe their pale colour to a thick coat of wax.

2. **Adaptations.**—Spinifex is well adapted to its peculiar habitat. The fixation of the plant in the loose subsoil, the supply of water under adverse circumstances, the constant battle against the sea-breeze to which it is exposed, as well as the taking advantage of this very breeze for the dispersal of its seeds, are shown to great perfection in this plant.

(a) Fixing in the soil.—Not only are various stocks connected with one another by strong runners, but they are also strongly fixed in the sand by numerous, deep roots which sometimes attain the length of ten feet and more.

(b) Supply of water.—The roots penetrate the sandy soil in every direction and reach a depth where the soil is always wet. Thus the plant is supplied with water. But that water is saltish. If much salt is absorbed, the growth of the plant is hindered (see Mangrove, page 53). The leaves, therefore, have not only a large water-storing tissue in them, which keeps the solution of salt at a low percentage, but they are also protected against a great loss of water through transpiration by means of that bluish coat of wax spreading over the epidermis.

(c) The plant and the wind.—Such a xerophilous structure is also necessitated by its constant exposure to strong breezes. For the same reason it must be strongly rooted in the loose sand, as seen above. As the plants are not tall, the wind can do them very little harm. On the contrary, they take advantage of the wind in letting it tear off the mature globular fruits from their dried stems. Rolling and dancing, the fruit-balls are carried away over the smooth sandy plain, dropping one seed after another on their way. By and by, the bristles between which the seeds lie, are worn away, and they are finally buried under the sand with the rest of the seeds.

7. **The Distribution of the Grasses.**—We have seen that the greater part of the food of man is derived from plants belonging to the family of the Grasses. It is, therefore, no matter of astonishment to learn that the Grasses, such as Rice, Wheat, Millets, cover the larger part of our cultivated areas. Grasses form the extensive meadows spread out over hills and dales; they inhabit the soft swamps as well as the beaten ground of the way-side; they thrive in the cool shade of jungles and on the scorched heath, on the sandy soil of coast tracts and on the rocky ground of mountains, in the arctic zone under snow and ice, as well as in the torrid zone under the parching heat of the sun; they form the extensive tracts known as the Prairies in North America, the Pampas and Llanos in South America, and the

Steppes in Europe and Asia: *of all the families in the vegetable kingdom the Grasses occupy the largest area of the fertile parts of the world.*

40. The Pine Family

(Coniferæ).

This family belongs to a group of families *(Gymnosperms)*, quite distinct from the families we have described heretofore *(Angiosperms)*. The distinctive feature of the Gymnosperms is in the seeds, which are not enclosed in an ovary, but are naked. The Conifers are inhabitants of the temperate and cold latitudes. Some of them are successfully planted on the high hills of India, *e. g.*, the Goa Cypress *(Cupressus glauca)*.

Woods of Casuarina *(Casuarina equisetifolia)* bear a striking resemblance to the Pine forests of northern regions. The tree does not, however, belong to the Gymnosperms, but to a distinct family *(Casuarineæ)* of the Angiosperms, allied to the Urticaceæ. It is leafless, the branchlets being green and cylindrical with sheaths of scales at the nodes. "The branches, when gently swayed by the wind, give out a sound like that of the sea on a beach, very pleasing to the ears of exiled islanders." The flowers are diœcious, sometimes monœcious. The tree is originally Australian, but now extensively cultivated in many parts of India as a remunerative fuel-tree.

DIVISION II.

FLOWERLESS PLANTS (Cryptogamæ).

These are plants without flowers. They multiply by spores, seed-like cells, that contain no separate germ like the seeds of the flowering plants.

The chief classes of this division are: the Ferns, plants with distinct stems and leaves and with vascular bundles; the Mosses, also with distinct stems and leaves, but without vascular bundles; the Fungi, without distinct stems and leaves.

41. Ferns (Filices).

Most of the Ferns live on shady and moist ground. One of the commonest on the West Coast of India is the creeping **Maidenhair Fern** (*Adiantum caudatum; Kan.* Čēḷukoṇḍi, Ānekivi), which we can find on every wall and rock during the monsoon.

1. **Stem.**—The stem of this Maidenhair Fern is a small, creeping root-stock, just on the surface of the soil with a bundle of small, fibrous roots. In the rainless season the plant withers down to its tiny stem in which its life is perpetuated.

2. Its **Leaves** or fronds, as they are called in ferns, are exceedingly *thin and tender.* The plants do not require a thick epidermis on their leaves to lessen the glare of the light or to reduce the action of evaporation; for they grow during the monsoon, when the sun is mostly screened behind clouds and there is always sufficient moisture for their growth, both in the soil and in the air.

The tenderness of the leaves, however, involves the danger of their being torn by the wind. This is somewhat avoided by the division of the leaves into a number of small segments: the leaves are *pinnate,* and the leaflets become smaller as they get farther away from the base of the petiole.

When the leaves grow out of the root-stock, they are *rolled up*, the very delicate parts of the young leaf-blade being inside and the strong and hairy leaf-stalk outside. When the latter unrolls itself, it first pushes aside the leaf-mould that may lie over it, and then gradually spreads its soft and thin blade to the light. A peculiarity of the fronds of this plant is that they end in a long tail which bends down, seeking the ground, and strikes root to produce a new young plant at its tip (fig. 165).

Fig. 165. — Creeping Maidenhair Fern *(Adiantum caudatum)*.

3. **Reproductive Organs.**—The leaves, developed at the end of the monsoon, have small growths on the margin of the lower side of the leaflets (fig. 166). They are green at first, but soon become brown and then look like withered

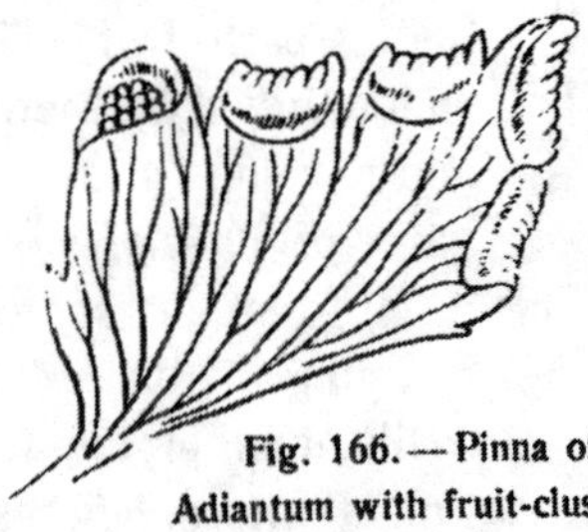

Fig. 166. — Pinna of Adiantum with fruit-clusters.

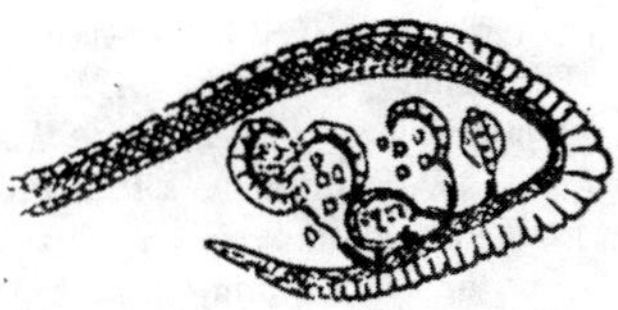

Fig. 167 — Section through the fruit-cluster of Adiantum showing the spore-capsules.

fringes. If we examine these growths, we find in them clusters of brown capsules, each about the size of a grain of sand. A lupe or, still more, a microscope will show that the capsules have the form of a biconvex lens, with a ring of thickened cells. If they are gently heated on a dry slide over a spirit-lamp and then quickly observed, we can notice how the ring of marginal cells is suddenly straightened and thus causes rupture of the thinnest point of the ring, and how it recovers its original curved position with a sudden jerk, by which the spores in the capsule are forcibly thrown out. This can only happen in dry weather when the wind can disperse the seeds. The spores are so minute and light that the wind can carry them far and wide to places at a distance from the mother-plant, where they settle and, under favourable circumstance, *i. e.,* at the beginning of the following monsoon, develop into fresh, little plants, which, however, are quite unlike the mother-plant. For they produce no stems and fronds, but have a minute, leaf-like body on which the essential organs of reproduction, male and female, grow. If we sow the spores on damp earth and keep them moist and sheltered from direct sunlight, they will germinate, and after a few weeks the surface of the soil will be found covered with small, green, flat bodies, each of which is an individual little fern, and develops those organs of reproduction. It is from the spores resulting from the fertilization of the female organs that the first type of the plant arises. *Ferns,* therefore, *pass through two successive states, one of them being a*

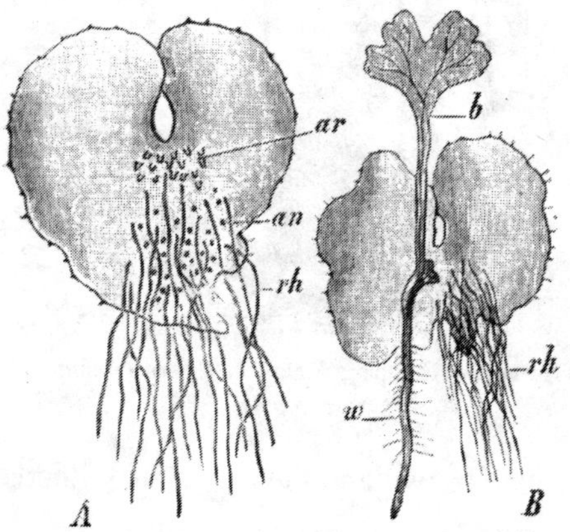

Fig. 168.—The flower-bearing type (called Prothallium) of a Fern. *A.* Prothallium from the lower side. *ar.* Female, *an.* Male organs. *rh.* Roots. *B.* Prothallium with a young fern growing from a fertilized ovule. *b.* First leaf, and *w.* root of the spore-bearing other type of the plant. (About 8 times enlarged.)

plant of considerable size and consisting of root, stem, and leaves with spores, and the other being very small and developing sexual organs from the seeds of which the first type again grows.

4. **Distribution and Classification.** This large order of plants is represented in nearly all parts of the earth, but in the

Fig. 169. — A group of tropical Ferns.

tropics, and there especially on islands and coast tracts, they grow with greatest luxuriance (fig. 169). The arrangement and structure of the fruit-clusters on the fronds are the characteristics by which the Ferns are classified. They are roundish, uncovered and scattered on the under surface of the fronds: in the genera *Polypodium* (compare the epiphytical *P. quercifolium*, fig. 170) and *Gleichenia*;—seated on the margin of the fronds and covered: in *Pteris*, *Adiantum*, and the creeping *Lygodium*;—cup-shaped and seated marginally: in *Davallia, Trichomanes,* and *Hymenophyllum;* — in continuous lines on each side of the mid-rib and covered: in *Blechnum;*—linear and scattered over the under surface and covered: in *Asplenium* and *Athyrium;* — globose, mostly on the back of leaf-nerves, with or without a cover: in *Aspidium* and *Nephrodium;*— sterile and fertile fronds are distinct: in *Acrostichum;* — the fruit-clusters cover only the upper parts of fertile fronds: in *Osmunda, Cyathea* and *Alsophila* are arborescent.

Fig. 170. — *Polypodium quercifolium*, epiphytical on the branch of a tree.

Much richer in Ferns was the earth in ages gone by, when many Ferns grew as large as trees, the atmosphere being then much damper and warmer than now. The falling stems of such tree-ferns were floated together by mighty streams, carried away to the sea, and buried under sand and mud. The remains of these plants, thus being shut out from the air, could not rot, but were slowly changed into *coal.* The impressions or casts of leaves and stems of Ferns can be distinguished in many pieces of coal even now (see fig. 171).

5. **Allied groups** of plants are the **Water-Ferns** *(Hydropterides)*, the **Selaginellas** (*Selaginellaceæ)*, and the **Club-Mosses** *(Lycopodiaceæ)*. A representative of the Water-ferns is *Marsilia*, a pretty creeping plant with quadrifoliate, folding leaves, common in rice-fields and shallow ponds. The Selaginellas are recognized by their many-branched stems covered by small alternating leaves. They spring up during the monsoon between Ferns and Mosses on earth walls. The Club-moss is likewise a denizen of moist soil. Selaginella and Club-moss have terminal clusters of spore-cases, whereas Marsilia has them hidden near the root.

Fig. 171. — A piece of coal with the impression of a Fern.

All these families, together with the Ferns, form one large division of the cryptogams, called *Pteridophyta*.

42. Mosses (Musci).

1. **Their Mode of Living.**—The Mosses live in large groups or colonies and form beautiful, green carpets on moist rocks or on humid and shady ground, and fantastic ornaments on the trunks and branches of trees.

The fronds of the Maidenhair Fern, as we have seen, are not in a position to live during the dry season. As they are unprotected against dryness, they have to wither and can continue their lives only through spores or through their root-stocks, which bring forth new fronds in the following rainy season. Mosses, however, are so constructed that their leaves simply fold and shrivel up during dry days and recover again after any refreshing shower of rain.

2. **Structure.**—A Moss plant consists of a short *stem*, decaying slowly at its lower end and continually growing at its upper, leafy end. It has no proper roots; but the lower end of the

stem is covered with a growth of brown, felt-like hairs (fig. 172), which penetrate the soil and act like root-hairs, absorbing water and food substances.

Fig. 172.—Hair-Moss (*Polytrichum commune*).

The *Leaves* at the upper end of the stem are of small size and simple form. They are often arranged in spirals. As they are very numerous and as Moss plants grow together in colonies, the rain water that falls on them is not only absorbed by the leaves in large quantities, but it is also retained in the spaces between single plants and between their leaves and stems.

Leaves and stems, when examined under a microscope, are found to be composed only of cells. They contain no vessels at all. Hence they are termed *Cellular Plants*. All the plants which we have hitherto noticed consist of cells and vessels, and are, therefore, called *Vascular Plants*.

If a group of Mosses, say of Hair-moss *(Polytrichum commune)*, is examined, we shall find

(a) some specimens with merely a bud, composed of young leaves;

(b) other specimens bearing, at their upper end, cup-like rosettes of leaves, which assume a bright-reddish colour and protect the minute reproductive organs (fig. 172, to the left);

(c) other specimens again bearing at their tips long-stalked spore-capsules, which we shall now study a little closer (fig. 172, to the right).

3. **The Spore-Capsule**—Both Ferns and Mosses are reproduced

by spores. The Ferns, as we have seen, form the spores in clusters of spore-cases, generally on the under-side of the leaf; the Mosses, however, throw up a stalked, urn-shaped body from the centre of the stem, in which the spores are produced. This little vessel is protected by a dry, fibrous hood *(calyptra)*, like the thatch of a hut (fig. 173, 1). When the spores are ripe, the hood is thrown off.

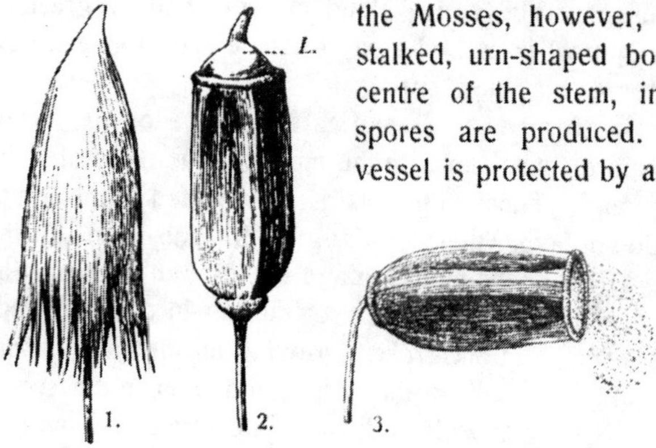

Fig. 173.— Capsule of *Polytrichum* (15 times enlarged).
1. Capsule with hood *(calyptra)*; 2. Capsule without it; *L.* Lid.
3. Lid fallen off; the wind sheds the spores.

The capsule, shut up by a small, pointed lid (2), now appears and soon places itself horizontally (3). The lid eventually drops also and discloses a pale-grey membrane, attached at its margin to the capsule by a number of tiny teeth (fig. 174). These teeth are very susceptible to moisture. When they get wet, they bend, and while bending press the membrane down, thereby completely shutting up the capsule with the spores. When dry, they stretch themselves

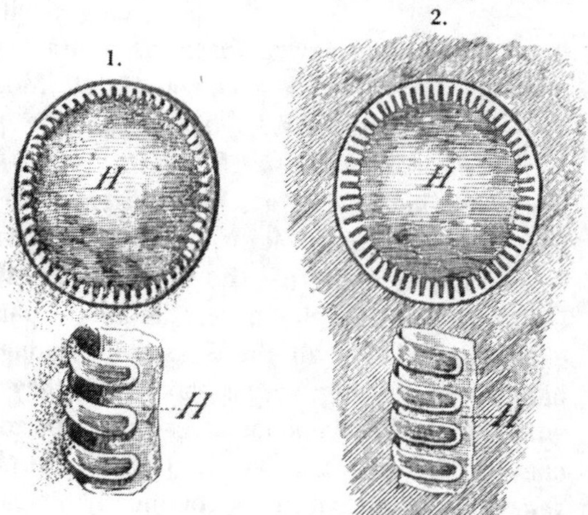

Fig. 174.— Upper face of capsule (30 times enlarged).
.1 When air dry: teeth loose, spores issuing between teeth and membrane *(H)*; 2. when damp: teeth tight, holes shut up.

upwards and lift the membrane over the brim of the vessel. If the wind then shakes the capsules on their tall, brown stalks, their contents pour out, and a cloud of yellowish or greenish powder (the spores) is carried away to some spot where a new colony of Mosses may now spring up.

4. **Distribution and Classification of Mosses.**—The Mosses are more conspicuous in the mountainous parts of India than in the plains. Those in the plains are dwarfed as the conditions there do not favour their regular growth. They come during the monsoon and wither with the cessation of rain. The commonest in the low lands are *Garckea,* growing on the earth, about half an inch high, and having the spore-capsule hidden in the terminal leaves of the stalk; *Hyophila,* growing chiefly on laterite rocks with very short stems and slender spore-capsules about one inch high; *Calymperes,* on the bark of trees, very short and reproducing itself by forming buds from the tips of its leaves; *Fissidens,* a minute plant with Fern-like prostrate leaves. The commonest hill Mosses are terrestrial *Polytrichum, Pilopogon, Pogonatum,* and the epiphytic *Thamnium, Penzigiella, Meteoriopsis, Papillaria,* etc. An allied group of plants are the Liverworts *(Hepaticæ)* which, together with the leafy Mosses *(Musci),* form the division *Bryophyta.*

Fig. 175.—*Garckea.*

5. **Importance of the Mosses in the Household of Nature.**—Mosses play a very important part in the household of Nature. First of all, the Mosses are among the *first settlers* on bare rocks. Being very small plants they content themselves with the smallest amount of earth, collected in the crevices or uneven parts of rocks; the old parts of the Mosses die off, form vegetable mould, and thus continually increase the little amount of humus in which they rooted at first. They thus gradually form a soil in which other, more highly developed plants can grow.

We have seen that Mosses generally grow not as single plants, but in groups, thus forming extensive, soft cushions. These cushions *absorb and retain the rain-water like sponges and give it off very slowly.* Thus the rivers and streams are kept supplied with water throughout the year, and the valleys through which they flow are rendered fertile. Mosses generally live together with trees and are abundantly found in forests, the shade and dampness of which are advantageous to their growth. Inversely they *help to keep the ground from drying up too soon,* and thus are useful to the trees which shelter them.

These facts help us to understand why the keeping up of forests is so beneficial to a country. They represent a reservoir of water, which is filled when it rains, and gives the water slowly off to the rivulets and streams, thus watering cultivated lands when there is no rain.

43. Mushrooms and Moulds (Fungi).

The Mushroom *(Agaricus).*

(*Kan.* Āḷimbe, Nāyikoḍe.)

1. **The Mushroom forming the Fructification of a Fungus.**—The Mushroom is a pale, soft substance, reminding one of a diminutive umbrella, with a short, stout stem and a large, horizontal head, at the under-side of which numerous vertical plates, radiating from the stem, can be distinguished. If the head of a mature Mushroom is laid on a sheet of white paper for a few hours, the paper will be covered with the dark powder of minute spores, produced between the laminated parts of the head. As long as the spores are not mature, those delicate parts are covered and protected from the effects of bad weather by a veil, which later on breaks away from the stem and leaves a ring-like scar on it (fig. 176).

2. **The Mycelium.**—The Mushroom grows from a dense network of white filaments, called the *mycelium.* The latter lives under the soil and grows continually, whereas the Mushrooms produced by it here and there are short-lived and perish as soon as they have strewn out their spores. The whole may, therefore, be compared, *e. g.,* to a Mango tree bearing numerous fruits which are dropped when they are ripe. *The mycelium is the*

fungus proper, and the Mushrooms are merely the fructifications. The plant lives under the soil; the fructifications, however, are raised above it in order that the wind may disperse the spores.

Fig. 176.—The poisonous Toadstool *(Agaricus muscarius).*

3. **Its Mode of Living.**—Like the roots of higher plants the filaments of the mycelium permeate the soil in every direction and draw their food from it. But, as we have seen on different occasions, the roots of plants take up only water and salts. These substances rise into the upper parts of the plant and are there, together with the carbon obtained from the carbonic acid gas of the air, converted into all those substances from which the body of a plant is built up. This work is done by the chlorophyll in the presence of sunlight.

But there is *not the least trace of chlorophyll* in a fungus. The Mushroom is, therefore, obliged to take its food up in a ready-made form; and it finds this in the decaying animal or vegetable matter of the soil in which it grows. It is a *saprophytic plant.*

Mushrooms can, therefore, grow only in places where such decaying matter is found. They do also not require any light for their growth, like plants with chlorophyll, and hence can be found in the darkest places.

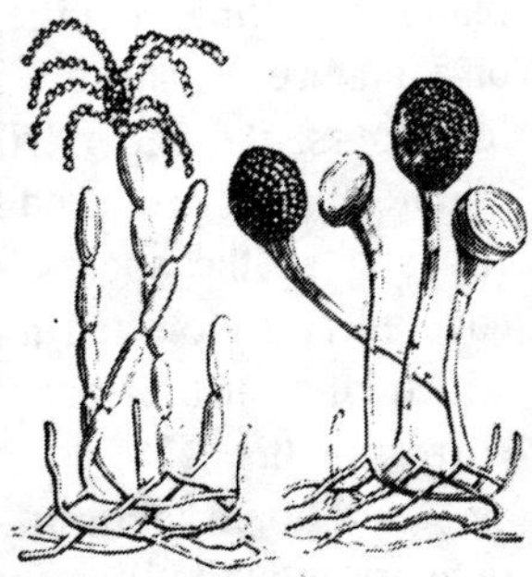

Fig. 177.—Mould with mycelium and fructifications (100 times enlarged).

4. **Importance of the Fungi in the Household of Nature.**—As we have already seen, the Mushrooms decay very soon and thus convert the animal or vegetable substances on which they grow, into nourishing matter for other plants. They may, therefore, be considered as helps to accelerate the process of decay, and are thus of great service to the animal and vegetable world.

On the other hand, many fungi are injurious to man, as they destroy large quantities of agricultural produce, timber, and other substances, when circumstances favour their development. Among these we mention the *Blights* which, as Bunt and Smut, destroy the grain in the ear of Paddy, and the *Moulds* which are so difficult to combat in the monsoon.

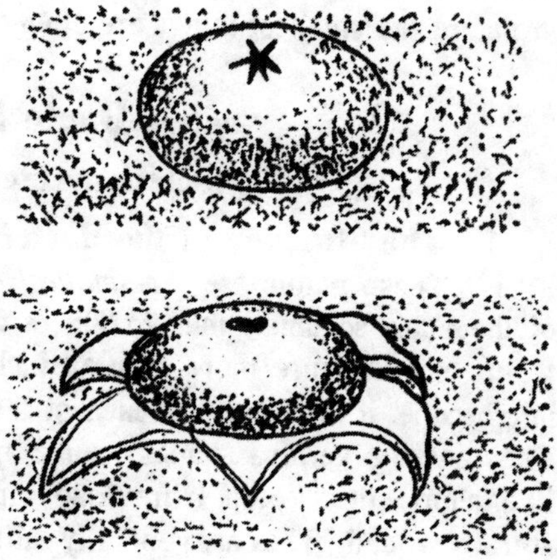

Fig. 178. — *Geaster.*

A common fungus in the plains of India is the *Geaster,* a fungus that has its fructification in a double cover. The outer cover bursts and spreads star-like, the inner one opens by a hole at its apex.

44. Lichens and Algæ.

The **Lichens** *(Lichenes)* represent a certain class of fungi which associate with another cryptogamic class, the **Algæ,** to have a common household (symbiosis). Like all green plants, the Algæ, which are green, are capable of forming the substances required for the building up of their tissue. Hence they live also by themselves on stones, walls, trunks of trees, etc. The fungus, however, is dependent on ready-made food. It takes such from the Algæ which it densely covers with its mycelium. In return, the fungus supplies its purveyors with raw food (*i. e.,* water and minerals dissolved in it), protects them from exsiccation and fastens them to the rock, or the trunk of a tree. The Lichens, which are commonly known by their grey colour and their crust-like, leafy or shrubby appearance, form with the Mosses (page 191) the very outposts of vegetation, growing at the expense of the atmosphere, and the moisture and dust which it bears to them. In course of time they make the hardest rock habitable for higher organisms, and thus show, once more, that in nature the smallest things are often of the greatest importance.

45. Bacteria

(Schizomycetes).

A. **The Structure of the Bacteria.—**

1. These plants are *the smallest organisms* known. Many of them are so small that 25,000 of them arranged end to end would not measure more than an inch. If examined under the microscope, it is found that each of them is made up of one cell. They assume various shapes: some appear like globules, some like short sticks, others are long and straight, others, again, spirally wound. They are generally called "bacteria" or "bacilli", words meaning sticks or little rods.

2. Under favourable conditions they *multiply by splitting in two.* And they multiply at such an enormous rate that, if the conditions remained favourable, one such minute thing would in the course of less than six days swell to a mass larger in bulk than the earth itself. This is, of course, never possible, for the food they require for their incessant multiplication would soon fail. But we can see from this example the tremendous rate at which they increase. If the conditions under which they live become unfavourable, they assume a shape that enables them to lie passive until the conditions again become favourable for the continuation of their life.

3. We have seen how small these plants are, and we can now understand that dry Bacteria can be easily whirled up by the wind and carried away thousands of miles. As invisible dust they are *present everywhere* in the atmosphere, and return to the earth when the air becomes calm.

B. **The Activity of the Bacteria.—**

1. The Bacteria, like the Fungi, lack chlorophyll; they are, therefore, dependent upon ready-made nourishment, animal or vegetable, which they obtain easiest in *decaying matter.*

(a) A simple experiment will, however, soon teach us that they do more than merely feed on decaying matter. We take two glass flasks with a little water, into which we put some animal substance. Then we close both flasks with a loose wad of cotton wool. The contents of one flask we leave undisturbed, but the other one we boil for some time, so that the Bacteria in it may be killed. The Bacteria cannot resist the temperature of boiling water any more than other organisms. After a day or two we shall find the contents of the unboiled flask begin to decay, that in the other one remaining unaltered. But if we remove the stopper from it, so that any Bacteria from the air can enter into it, decay sets in here likewise. This shows that the Bacteria do not only live on decaying substances, but that they are also *the cause of decay.* In other words, *there would be no decay on the earth without the Bacteria.*

(b) Suppose the latter were the case. Millions of corpses of animals and plants would cover the earth without decaying. This would result in the destruction of vegetable life, as plants would not find the required food in the soil which is produced by such decay. And in consequence of the destruction of vegetable life also, the animals could no longer exist. It is the Bacteria which cause decomposition and thus are the chief cause of the *continuous cycle of matter in nature.*

(c) In this connection we may consider a very important thing referring to agriculture. Each time the crop is removed from a field, a large quantity of nitrogen, deposited chiefly in the seeds, is taken away from the field together with other nourishing elements of the soil. The plants are not able to absorb from the air the nitrogen which they cannot do without. It must, therefore, be restored somehow, and this is done by manuring the field. If, however, fresh manure is used, plants will not grow well, and often die. The albuminoids contained in fresh manure must be rendered soluble in order to be of use to plants. This is done by the Bacteria in the soil by decomposing them. *Manure is thus, by the agency of Bacteria, transformed into such a state that it can be used by the plants as nourishment.*

(d) Plants are, as has been just remarked, not able to take their supply of nitrogen from the air which has such an abundant quantity of it (about 79%); certain Bacteria form, however, an exception. They grow as parasites on the roots of the Leguminosæ and form nodules on them (*cf.* Pea, page 35). These nodules, when squeezed, throw out a sticky fluid which really consists of innumerable Bacteria that can be readily recognized under a powerful microscope. *The Bacteria that form these nodules, are able to derive their nitrogenous food from the air*, which higher vegetation is unable to do. The larger the quantity of root-nodules, the greater the amount of nourishment derived from the air and stored in the soil. The advantage of growing Pulses, Sunn-Hemp or other Leguminosæ, to recoup the land, is explained by the peculiarity of these plants.

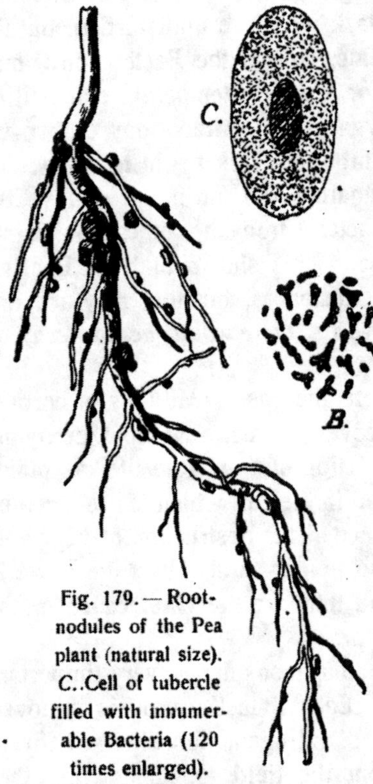

Fig. 179.—Root-nodules of the Pea plant (natural size). *C.* Cell of tubercle filled with innumerable Bacteria (120 times enlarged). *B.* Bacteria (800 times enlarged).

2. Certain other Bacteria produce, in the substances on which they live, a change which is not called decay, but *fermentation* (*cf.* page 22). Again, if toddy or wine is allowed to stand open for a few days, it becomes sour. This is also due to the action of some Bacteria. Similarly it is the Bacteria which turn milk sour or spoil boiled rice and vegetables. By the action of Bacteria the fibres of Sunn-Hemp *(Crotalaria)* are loosened; and also the peculiar flavour of Cocoa and Tobacco is due to the influence of these little organisms which cause fermentation.

3. Plants without chlorophyll find suitable nourishment not only in decaying matter, but also in *living organisms*. It is no wonder that we should, therefore, find numerous *parasites* among the Bacteria. They penetrate the bodies of animals and men, multiply there at a rapid rate, and produce a number of deadly diseases. Of these diseases we shall mention here only these few: — Consumption, of which one-seventh of all men die; typhoid fever, diphtheria, pneumonia, and influenza, which also every year destroy a great number of men in the prime of their life, and cholera and plague, which are the most terrible scourges to which a country can be subjected.

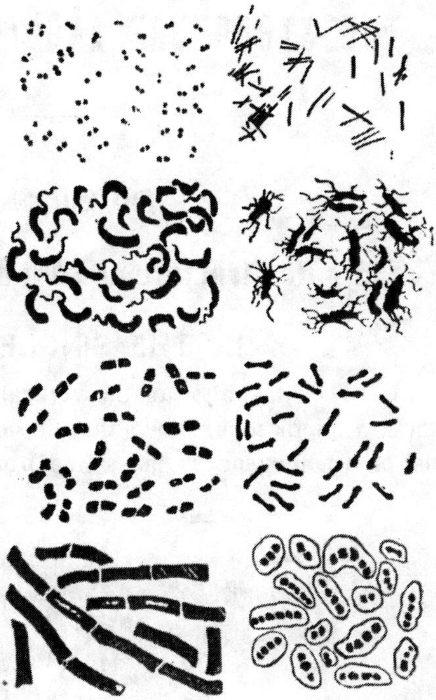

Fig. 180. — Bacteria of tuberculosis, typhus, diphtheria, pneumonia, influenza, cholera, plague, etc.

It is right that we should learn to know *how to meet these powerful enemies*. One of the means to keep these "omnipresent" microbes away from us is the greatest *cleanliness*. This holds good of the vessels in which we prepare and preserve our food, of our houses and their surroundings, of our garments, and, above all, of our own bodies. As shown in the above experiment, Bacteria perish at the temperature of *boiling water*. This supplies us with a means of keeping good, for some time at least, a number of food substances, such as meat, fruit, vegetables, milk, etc., which would otherwise soon be spoiled. From immemorial times, men have also been using salt to preserve meat and to pickle fruits, sugar to candy them, and smoke to preserve meat. Besides, there are some drugs, called *antiseptics*, which also destroy Bacteria or prevent their action.

SECOND PART

THE STRUCTURE AND LIFE OF PLANTS

DIVISION I.

The Minute Structure and Vital Processes of Plants.

I.—THE SINGLE CELL

1. The illustration of the vertical section of a leaf, as seen under a microscope, shows that the leaf of a plant does not consist of a homogeneous mass like iron or glass, but of several

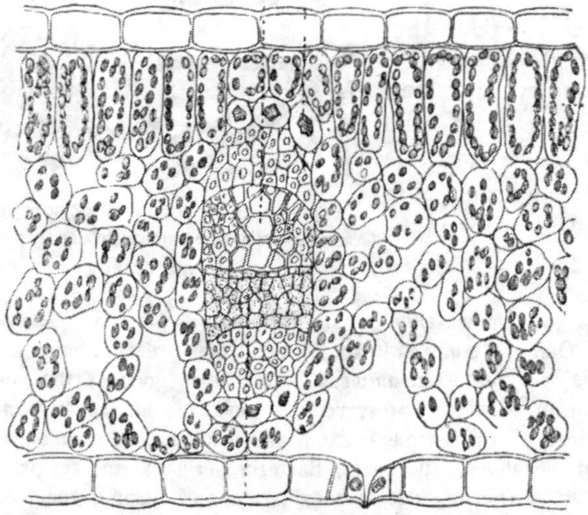

Fig. 181.—Transverse section of a leaf, to show its composition of cells.

small parts, called **Cells,** which like the stones of a wall, constitute the whole. Like the leaf, the other parts of a plant, *viz.*, root, stem, flower, and fruit, are all composed of cells.

2. Most plants are thus **composed of numerous Cells.** But many of the lower classes *(e. g., Bacteria)* consist of a **single cell** each. These plants are generally very small.

3. The **Size** and **Shape** of the cells vary greatly. In the Bacteria the length of one is less than ·001 cm, in the fibre of Flax and the hair of the Cotton-seed it extends over 4·cm. Their shape may be spherical (as in the pollen grains), or cubical, or brick-shaped (many epidermis cells) or cylindric, or tapering towards both ends like a torpedo (as the bast cells of most plants).

4. In some cases a number of cells standing one over the other in a row have the partitions which separate them removed, and thus form a tube or a **Vessel,** open all through.

Plants, then, are built up of cells, or of cells and vessels, the latter originating from cells.

5. The cell consists of a cell-wall and the cell-contents, which when young is called **Protoplasm.** This is a viscid, nitrogenous substance, capable of absorbing moisture, of expanding, of forming fresh cells by division, and of motion: it is *endowed with life.* The cells are the workshops in which all the secret and wonderful operations of the plant-life are carried on.

6. By the activity of the protoplasm certain substances are produced, which are of great importance to the life of the plant. Some of these are green granules, called **Chlorophyll,** which give the plants their green hue. The chlorophyll-granules have the power of forming starch under the action of sunlight out of carbonic acid gas and water absorbed by the plant.

7. Besides, the cells contain a fluid, called the **Cell-Sap,** in which sugar, acids (citron), salt, and other substances are dissolved. This sap rises from cell to cell, permeating through the cell-walls from the root to the top of the tree. The protoplasm forms various substances out of the cell-sap, which are either passed along to any point of activity where their presence is necessary for the growth of the plant; or they are stored up in the tissues of the plant for future use. In this way starch, oily and fatty matters, and grains of protein are formed and stored.

2.—THE ORGANIC STRUCTURE

1. Plants which consist of *single cells,* like the Bacteria, can be compared to single men who live for themselves and have to do everything alone, such as gathering of food, building of houses to live in, defending themselves against enemies and so on. So single cells have to do all the functions essential for life: they have to absorb their food and digest it, to excrete waste substances, to guard against adverse circumstances, and to reproduce their kind.

2. Plants consisting of *various cells*, however, are like a state, in which the different labours conducing to the welfare of the community are divided. As in the state certain individuals (farmers) are occupied in obtaining food for all, others (craftsmen) in supplying the public with houses and clothes, others (merchants) in the distribution and circulation of food and articles, and others (soldiers) in the maintenance of order and in the defence of the common-weal, so the various cells of a plant are assigned different functions and form a well-organised state.

3. They are also specially fitted for their several special purposes, and groups of them thus form *organs* for the vital operations of the plant.

These organs are not equally developed in all plants. Some have a higher, and some a lower organisation. We shall now study the various organs of the plant, as we find them in their leaves, roots, stems, flowers, and fruits.

DIVISION II.

The Structure and Vital Processes of the Parts of Plants.

I.—THE LEAF

A. The Outer Structure.

1. **The Parts of a Leaf.**—The chief parts of a leaf are:

(a) the **blade,** which provides a large surface exposed to the action of light and air, so as to enable the plant to evaporate its water and to gather carbonic acid gas;

(b) the **stalk** or **petiole,** which places the blade in the most suitable position with regard to the light (see Cucumber, page 75) and protects it from damage by wind or rain (see Mango, p. 25).

Some leaves have no petioles, *e.g.*, Ixora, and are then called *sessile.* The petiole may have appendages on it quite distinct from the stem, as in the Orange (p. 19): it is then called *winged.* If the blade of the leaf runs down into the stem, as in Sphæranthus, the leaf is termed *decurrent.* When the petiole is not attached to the base of the leaf but to its centre as in Castor or Taro (p. 147), it is *peltate.*

In many leaves (Horse Gram) the base of the petiole is thickened and flexible. Such a leaf is called *pulvinate.*

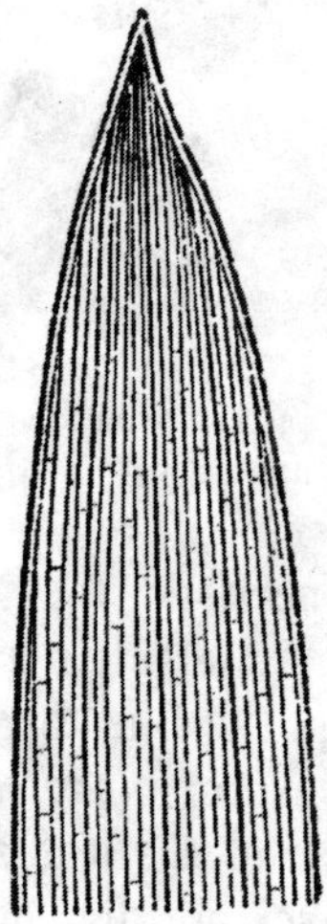

Fig. 182. — Tip of a Grass blade with parallel veins (4 times enlarged).

(c) The **sheath** is an expansion of the petiole surrounding the stem and protecting the tender parts beneath it. It is common in Grasses and other monocotyledons.

In the place of a sheath we find persistent or deciduous appendages, the **stipules,** at the base of the leaf-stalk in some plants. They are characteristic of certain families, *e. g.*, the Malvaceæ, Leguminosæ, Rubiaceæ, Euphorbiaceæ. Such stipules may also be modified and appear as thorns, as for instance in the Acacia.

2. The Blades of Leaves exhibit various **Characteristics:**

(a) As to the **arrangement of their ribs** or veins. If we examine the leaves of Grasses or Lilies, we shall find that the veins run parallel or almost parallel to the mid-rib in one direction from the stalk to the tip. The venation of such leaves is said to be *parallel.* The veins in most dicotyledons, such as the Teak and Banyan, form a kind of network; they are called *net-veined* (reticulated, fig. 184).

Fig. 183. — Stipulate leaf of Pea.

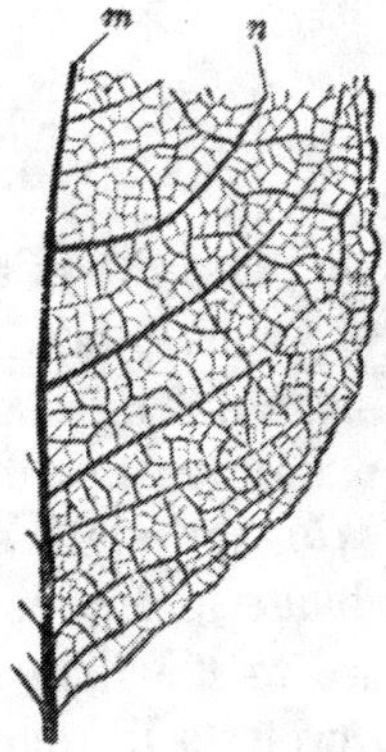

Fig. 184. — Piece of the leaf of a dicotyledonous tree with reticulated veins (reduced).

If the side nerves branch off from the mid-rib alternately or opposite like the hairs of a feather, we call the leaves feather-veined or *pinnately veined* (Adhatoda). If they radiate from the base of the leaf as the fingers from the palm of the hand, the leaves are digitately or *palmately veined* (Castor).

(b) As to their **general form:** Some are *linear* (Hemidesmus), some *oblong* (Mango), some *elliptic* (Vinca), and some *orbicular* (Jujube). If the broadest part of the blade is below its centre, the leaf may be *lanceolate* (Gloriosa), or *ovate* (Banyan); if, however, the blade is broader near its apex, it is *oblanceolate* (Crotalaria retusa), or *obovate* (Cashew).

(c) As to their **base:** *acute* (Mango), *narrowed* (Teak), *cordate* (Betel), *reniform* (Hydrocotyle), *auricled* (Calotropis), *sagittate* (Limnophytum), *hastate* (Typhonium).

(d) As to their **apex:** *acuminate* (Peepul), *acute* (Mango), *obtuse* (Cashew), *retuse* (Crotalaria retusa), *mucronate* (Cæsalpinia).

(e) As to their **margin:** *entire* (Jack), *dentate* (Clerodendron), *serrate* (Rose), *crenate* (Hydrocotyle), *sinuate* (Momordica charantia), *angled* (Luffa).

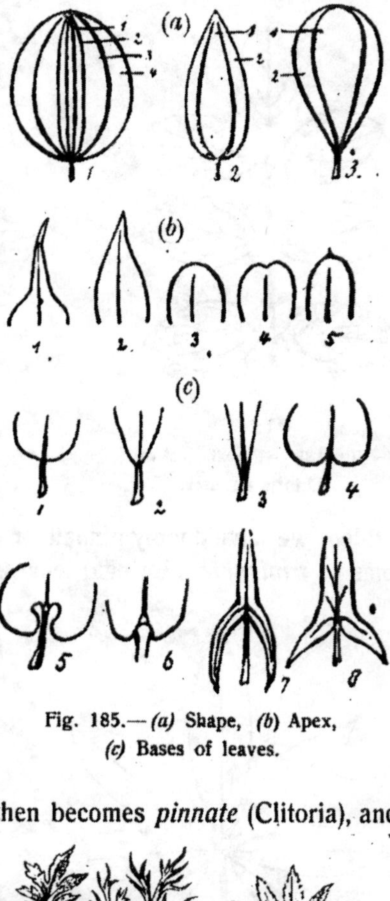

Fig. 185.—*(a)* Shape, *(b)* Apex, *(c)* Bases of leaves.

(f) If the incisions are deep, but do not extend half-way to the mid-rib we call them *lobed* (Urena lobata), if they go more than half-way: *pinnatifid,* or palmatifid (Jatropha multifida), and when they go almost to the mid-rib: pinnately or palmately *partite* (Manihot utilissima).

(g) If the incisions are so deep that the blade is divided into several distinct leaflets, we speak of **compound** leaves in contradistinction to simple leaves. The pinnatifid leaf then becomes *pinnate* (Clitoria), and

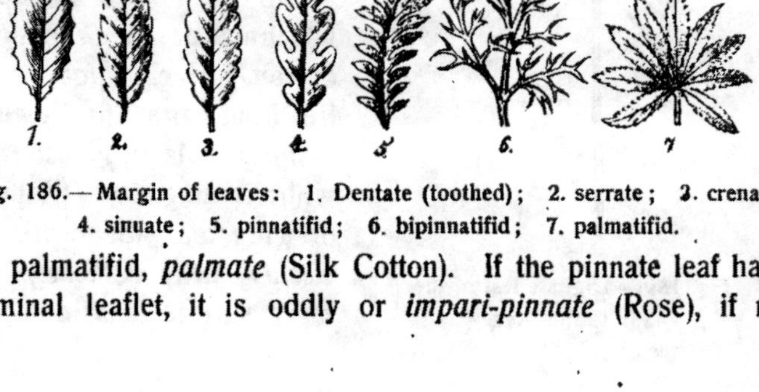

Fig. 186.—Margin of leaves: 1. Dentate (toothed); 2. serrate; 3. crenate; 4. sinuate; 5. pinnatifid; 6. bipinnatifid; 7. palmatifid.

the palmatifid, *palmate* (Silk Cotton). If the pinnate leaf has a terminal leaflet, it is oddly or *impari-pinnate* (Rose), if not, evenly or *pari-pinnate* (Pea). Are the leaflets (pinnæ) again

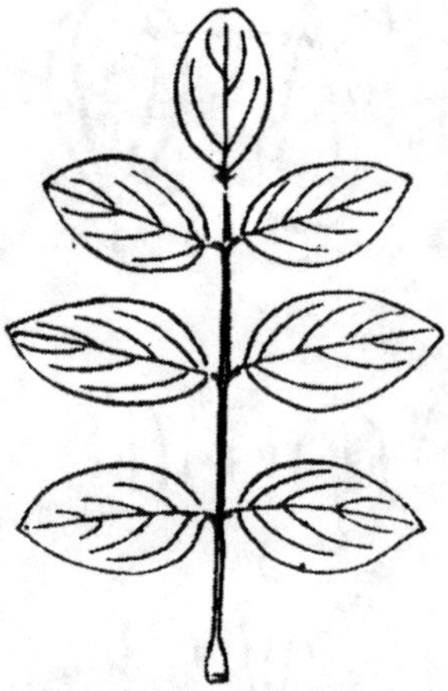

Fig. 187.—Pinnate leaf of Clitoria ternatea.

Fig. 188.—Trifoliate leaves of Horse Gram.

divided, we have doubly pinnate or *bipinnate* (Acacia) and three times or *tripinnate* (Moringa) leaves.

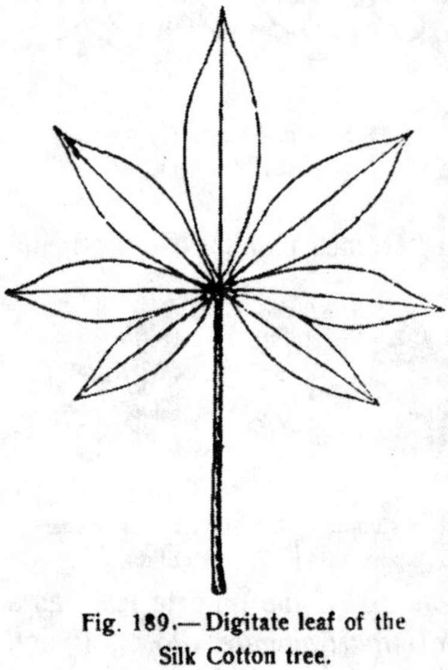

Fig. 189.—Digitate leaf of the Silk Cotton tree.

An interruptedly pinnate leaf is found in the Potato and Tomato.

A compound leaf of three leaflets may be either pinnately trifoliate (Horse Gram), or palmately trifoliate (Bael Tree).

(h) As to their **surface**: which may be *smooth* (Mango, or *rough* (Jack), *glabrous* = free from hairs (Shoeflower), or *hairy* (Tulasi); *glaucous* = with a coating of wax (Poppy) or *glandular* (Sundew), or *viscid* = sticky (Cleome). A hairy surface is said to be

pubescent, if the hair is soft and downy (Gingily), *tomentose* if rough and woolly (Ærna) lanata), *hispid* if the hairs are stiff (Ficus asperrima).

(i) As to their **texture:** *succulent* = fleshy (Bryophyllum), or *coriaceous* = leathery (Mango), or *membraneous* = thin and flexible (Rose), or *scarious* = thin and dry (stipules of Begonia).

3. The **Arrangement of the Leaves on the Stem** is such as to prevent any interference with one another, and thus to

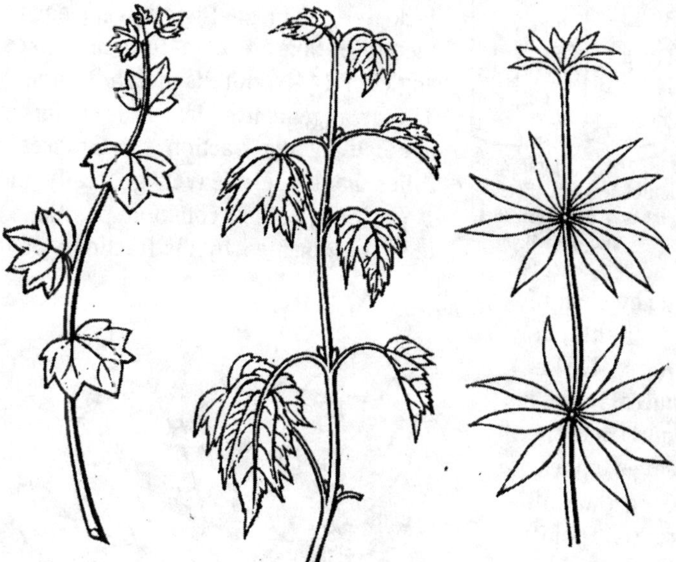

1. 2. 3.

Fig. 190.— Arrangement of leaves: 1. alternate; 2. decussate; 3. whorled.

allow free access to both light and air. They are either *alternate* when they originate singly from the nodes (shoeflower), or *opposite* when the leaves are in pairs all up the stem (Tulasi), or *whorled* if three or more leaves are at a node (Alstonia; *Kan.* Hālēmara). Opposite leaves are termed *decussate* when each pair of leaves is at right angles to the next pair (Labiatæ, Acanthaceæ, Rubiaceæ). Leaves are said to be *bifarious* when the leaves are all on the same side of the stem, left and right (Custard Apple).

If we repeat the experiment with the thread, described on page 65, we shall see that those seemingly irregular leaves are arranged on the stem in *spirals*. With the help of the thread it is also easy to find out how many turns are necessary to get to a leaf which is exactly above the first, and which of the number of leaves it is. Thus we count five leaves on two turns in the case of the Jack tree, or in the Shoeflower; each leaf, therefore, takes $\frac{2}{5}$ of a turn or makes an angle of 144° with its neighbouring leaf. The arrangement of leaves is in this case denoted by the fraction $\frac{2}{5}$. Grasses and Lilies have their leaves generally in the $\frac{1}{2}$ position; other common positions are those represented by the fractions $\frac{1}{3}$ and $\frac{3}{8}$.

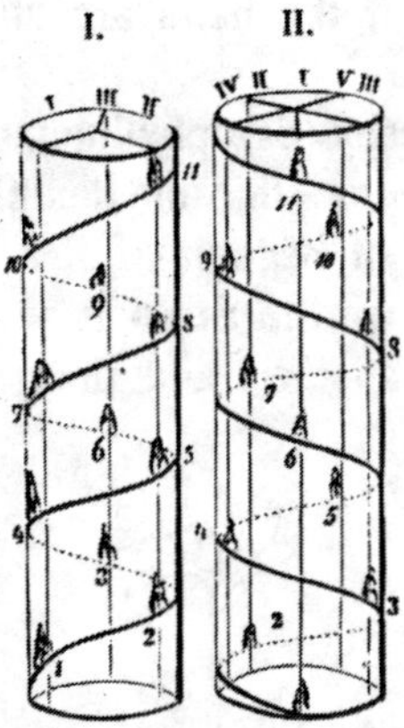

Fig. 191.—Spiral insertion of leaves. Each leaf takes $\frac{1}{3}$ of a turn in I, and $\frac{2}{5}$ in II.

The stem of Elephantopus (*Kan.* Nelamuččala) is so short that the leaves appear to grow all from one point. Such leaves are also arranged in spirals; they form what is called a *radical leaf-rosette* and show very plainly that the leaves so arranged allow one another their due share of light.

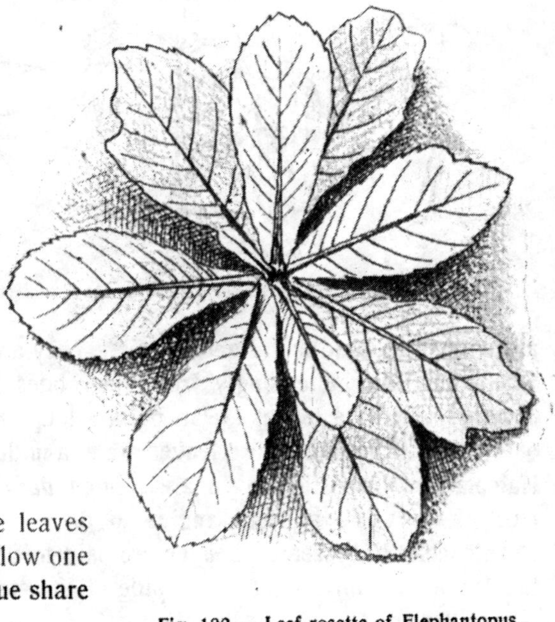

Fig. 192.—Leaf-rosette of Elephantopus.

4. The conventional way of describing leaves is to begin with their arrangement on the stem, then to continue with the description of the leaf-stalk and the stipules, then to state whether it is simple or compound and to proceed to its general shape, its base and apex (if distinctive), its margin, surface, texture, venation and colour.

5. **Metamorphosed Leaves.**—The leaves of one and the same plant are by no means always all of the same shape. As they answer different purposes, they are modified in shape. The main function of the foliage leaves is, as we shall see hereafter, the absorption of gaseous food from the air. Now, the *seed-leaves,* to begin with the first leafy structure of a plant, have to supply food to the young plantlet, and such food may be stored in the seed-leaves themselves (Horse Gram, p. 30) or in a separate tissue of the seed, the endosperm, which the seed-leaves have to absorb (Castor, p. 123, Maize, p. 169, Cocoanut, p. 139). Their shape, is therefore, such as to suit these special functions and often widely differs from the shape of the foliage leaf of the same plant.

Then, there are the leaves of the lower parts of the stem which are often rudimentary. Thus, for instance, in the Bamboo they consist only of the sheaths of the foliage leaves. In the underground stems of the Onion, the Potato, the Canna and similar plants they are *scaly*. We find such scaly leaves in the resting buds of woody stems as well, and here they represent the leaves at the bases of fresh shoots.

Again, the *bracts,* leaves found on the flower stalk, must be considered modified leaves. They differ from foliage leaves in size, shape and sometimes in colour also (see Adhatoda, p. 101, Sunflower, p. 64, Bougainvillea, p. 128).

The modification of leaves finds another expression in those plants, in which two kinds of foliage leaves occur on the same stem, as for instance in Emilia sonchifolia, where the lower foliage leaves forming a leaf-rosette are stalked and round, whereas those higher up on the stem are sessile and lyrate with auricled bases.

Specialized forms of leaves are

(*a*) the *tendrils* in the Pea, in Gloriosa, in Smilax and in Cucumber, growths that have to assist the stem in climbing,

(*b*) the *thorns* in Acacia, or in Jujube, these growths being metamorphosed stipules, and serving as instruments of defence to the plant against the attack of herbivorous animals,

(*c*) the *bladders* of Utricularia, which are traps by means of which the plant catches small animals on which it feeds.

B. The Work done by the Leaf.

1. Transpiration or Evaporation of Water.

(*a*) **The Fact that Plants evaporate Water** can be proved by an experiment. Place some fresh twigs of a plant under a bell-jar in the sunshine. After a short time we shall find a deposit of moisture on the inner side of the glass. Another bell-jar with no plants under it, similarly placed in the sunshine, has no such deposit of moisture. From this we draw the conclusion that plants transpire or evaporate water in the form of water-vapour.

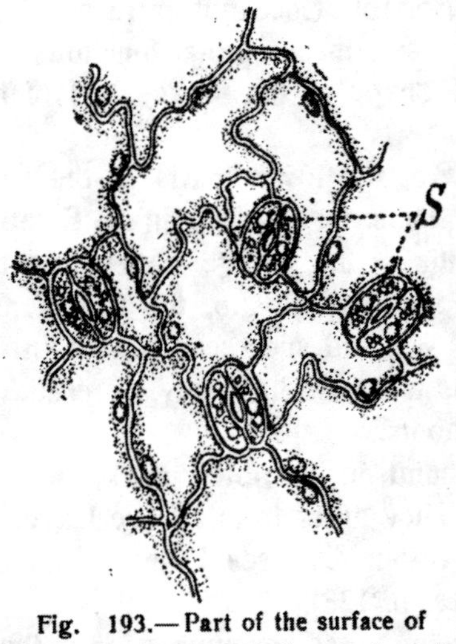

Fig. 193.—Part of the surface of a leaf. *S.* Stoma (200 times enlarged).

(*b*) **How can this Transpiration take place?**—A careful examination of the surface of leaves shows that there are many tiny openings which, looked at through a microscope (fig. 193), appear like little mouths, and are therefore called *stomata* (from the Greek *stoma*, mouth). Some leaves have them on both sides, but most only on the under side. (The Lotus plant cannot but have them only on the upper side. Why? Compare page 3.) These openings lead to hollow spaces, called *air-chambers,* in the interior of the leaves (see fig. 181 on page 198). The stomata are the "gates" of these cavities, through which a great deal of the water in the plant escapes as vapour. In a lesser degree transpiration also takes place through the walls of the cells of the outer skin, called *epidermis*.

(c) **Importance of Transpiration.**—This process is of the greatest importance to the life of the plant. It is generally known that plants suck up water and mineral food dissolved in water by means of their roots. This liquid food is carried up through the stem and the branches to the leaves, where, as we shall see later on, certain materials, which the plant requires for its growth, are formed. By a simple experiment we are able to find out which way the water-current takes in a leaf: we place a white flower, *e. g.,* a Eucharis Lily, with its stalk in water stained red with eosin. In a short time the red liquid will be seen rising in the white petals along the veins.

All the water that comes up is not required for the growth of the plant. When it has done its work as carrier of the mineral food from the soil to the leaves, it is passed off as vapour, making room for further supplies from the soil. The transpiration of water from the leaves thus *acts like a suction pump: it is always drawing up fresh supplies of water and food from the roots.*

(d) **Amount of Transpiration.**—We can ascertain the quantity of water evaporated by a certain plant in a given time, by means of a simple experiment. We put the stem of a twig with leaves into a tumbler of water, cover the surface of the water with a coat of oil and place the whole on a balance. After a few hours we shall notice a considerable loss of weight, due only to the evaporation of water through the leaves of the plant. Thus, for instance, it was found out that a Sunflower plant gives off a quart of fluid in twenty-four hours. Some large trees give off a thousand pounds of water in a single hot day. If the height of such a tree be a hundred feet, the work done by that tree amounts to the same as that of carrying some three hundred ordinary buckets of water up a stair ten feet high. A piece of land covered with trees brings an enormous amount of water up from the depth of the earth to the atmosphere. From this we may now easily understand, how important woods are for the fertility of a country, and how disastrous it is to destroy

forests. In each plant an invisible stream of water rises, as it were, from the ground to increase the ocean of the atmosphere, afterwards to come down again to the earth as rain.

The amount of evaporation varies with certain circumstances. In the first place, it depends on the temperature: the warmer the air is and the hotter the leaves become under the rays of the sun, the more rapid will be the action of transpiration. Secondly, when the wind blows and carries away the air round the leaves, which are saturated with vapour, bringing dry and thirsty air ready to take in vapour from them, the amount of evaporation will naturally be greater than when there is no wind. Thirdly, clothes dry much more slowly during the monsoon than when the dry land-wind blows. For the same reason plants will evaporate much more water in dry weather than in the monsoon. When the air is saturated with moisture, *e. g.,* in the cool morning of a monsoon day, no water can be evaporated from the leaves. Some plants, as Taro *(Colocasia),* Maize, Bamboo, are able to press the water out from little pores at the tips and edges of their leaves. These drops of water hanging from the tips of the leaves should not be mistaken for dew-drops.

Fig. 194.—A drop of water oozing from the leaf of *Colocasia.*

(e) **Helps to promote Transpiration.**—As the evaporation of water through the pores of the leaves is of such great importance to the plants, we find many arrangements in the plant-world by which this process is enhanced when likely to prove useful to the plant.

Plants that grow on shady and moist places have *large leaves,* as a rule, and generally very numerous stomata. Their leaves are, besides, very tender, that is to say, their *epidermis is so thin that water can pass* not only through the pores (stomata), but also *through the walls of the surface cells.*

We can sometimes see dark spots or blotches on the leaves of some plants, such as some Aroideæ or Turmeric. *By virtue of* these *dark-coloured spots the leaves are enabled to absorb more heat* than if they were green throughout. (A dark coat feels warmer than a bright one.) Similarly, the leaves of the Lotus plant are coloured dark-purple on their under side.

Many plants, like the Bean and almost all Leguminosæ, have the curious habit of *folding their leaves at night.* This, again, is an ingenious device to prevent dew from covering them and thus choking the pores, so that the action of transpiration may not suffer (see page 34).

Fig. 195.—The Silk Cotton tree in the dry season: leafless but fruit-bearing.

Other plants, like the Opium Poppy and Colocasia, possess a *bloomy coat of a waxy substance* on their leaves so that they are not wetted, and the process of evaporation may not be interrupted.

Most leaves *taper into a point* instead of having a blunt end. This enables them to become dry soon after the rain ceases, as the raindrops fall easily to the ground from such tapers.

Plants that grow under favourable conditions of water-supply are termed *Hygrophytes*.

(f) **Means to check too much Transpiration.**—Too much transpiration, *i. e.,* too great a loss of water is, on the other hand, dangerous for the life of a plant. Plants, which have to live on very dry soil, or are exposed to the scorching heat of the sun or to the parching influence of dry winds, or which for many months must do without a drop of rain, must needs have some means of reducing the action of transpiration.

Above all, we notice that the leaves of such plants have a very *limited surface (cf.* Leucas, page 104), and are sometimes reduced to mere spines or scales (Cactus, page 58; Casuarina, page 181). Some plants drop their leaves entirely during the dry season (Teak, page 112; Silk Cotton tree, fig. 195) to prevent loss of moisture by evaporation.

Another kind of protection from the scorching heat of the sun is the *vertical position* of leaves, as is seen in many trees of Australia, of which the Eucalyptus tree is one. Instead of holding their leaves flatly or horizontally, as trees generally do, so as to catch every ray of sunlight, they avoid the heat as much as possible by holding them edgeways to the light. Similarly some other plants, like Oxalis (*Kan.* Puḷḷampuṇise; *Mal.* Puḷiyāral; *San.* Čukrikā) fold their leaves at midday when the sun shines hottest.

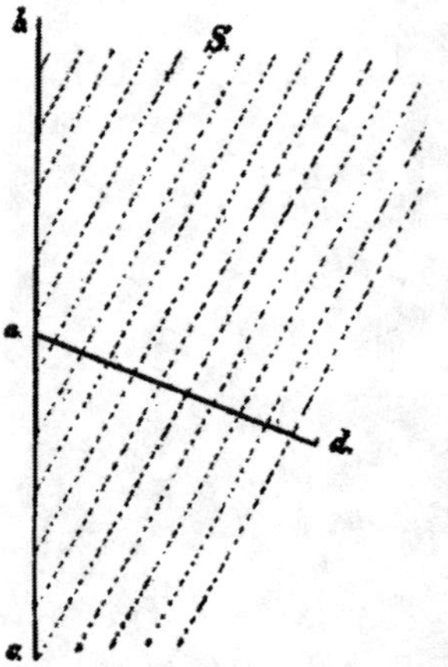

Fig. 196. — Solar rays striking the line *a d* at right angles and heating it much more than the lines *a b* or *a c* which lie more in the direction of the sun's rays.

Various plants that live on dry and rocky soil, store up in their *fleshy stems and leaves* every drop of water they can get when rain falls, and live sparingly on it during the long periods of drought which may last for three-quarters of the year. Such plants are the Bryophyllum (*Kan.* Kāḍu-basale), the Agave (*Kan.* Ānekattāḷi), and many Spurges and Cactuses. A number of Orchids make the same provision for long periods of drought in using their enlarged stems or pseudo-bulbs as storehouses of nutriment and water, upon which the plants feed in times of need.

And not only are these plants thus enabled to store up water against the time of drought, they also keep this precious fluid under a *thick, leathery skin,* through which only very little moisture can escape.

Another very common means of protection against too great a loss of moisture by transpiration is the *hairy covering* of their leaves. The hairs keep a layer of quiet air within their spaces and thus prevent the access of new, unsaturated air which would dry up the leaves in a short time. (See Anaphalis, p. 71).

Many leaves also are *shiny,* and reflect a great deal of the heat which would otherwise raise their temperature, and so increase the activity of evaporation (*cf.* Mango, page 25).

And, last but not least, most plants have the wonderful power of *closing their stomata* as soon as the amount of water coming up from the roots becomes scarce.

Plants with such contrivances to keep down the loss of water to a minimum are termed *xerophytes.*

2. Assimilation of Food.

I. Absorption of Mineral Salts by the Roots.

In the preceding chapter we have seen that plants need a constant supply of water. The water taken up by the roots contains substances from the earth dissolved in it. Which are these substances? If we burn a plant carefully, an ash remains. The minerals constituting the ash, must have been the substances absorbed by the plant. Chemists tell us that the principal minerals found in the ash of plants are *sulphur, phosphorus, potash* or *soda, lime, silex, magnesium* and *iron.* Other substances that formed part of the body of the plant, but were dissipated by the heat, are *carbon, water,* and *nitrogen.*

All the substances named above, excepting carbon, form the mineral food absorbed by the root. We prepare a solution of

1 gram of potassium nitrate, 0·5 gram of sodium chloride, 0·5 gram of calcium sulphate, 0·5 gram of magnesium sulphate and 0·5 gram of calcium phosphate, in 1000 grams of distilled water, and add one or two drops of iron chloride. If a seedling, say of Pea, is nursed in this food-solution, it will be found that it grows healthily and even develops flowers and seeds. But if one or the other of the minerals is omitted, the growth will be stunted and dwarfed. If, for instance, there be no iron in the solution, the leaves and the stem of the plant will become pale yellow, but they turn green when a drop of liquid containing iron is added to the solution.

The minerals thus absorbed with the water rise to the leaves, and are there assimilated to form elaborate food for the formation of new structures.

II. Absorption of Carbon by the Leaves.

(a) **Source of Carbon in Carbonic Acid.**—The greater part of the plant-body consists of carbon (charcoal is carbon). The plant nourished in the food-solution, as described above, could not derive it from the water. It must, therefore, have had some other source of carbon supply.

Atmospheric air always contains, more or less, (0·03—0·04%) carbonic acid gas, a compound of carbon and oxygen, and given off from the lungs of animals and men, and from the burning of wood and coal. This is the source of the carbon supply. The leaves absorb the gas, and because there is so little of it, each tree needs to spread out an immense amount of foliage, so that it may drink in all the carbonic acid gas that can possibly be obtained.

Some very instructive evidence is furnished by a small aquarium. If animals, especially fish, alone be kept, it will be found necessary to renew the water daily, or the animals will soon die. If, however, some water-plants are introduced, the same water may be kept in for months, and the animals will continue healthy, thus showing that animals soon make the water fatal to themselves, and that plants restore and maintain the balance, evidently taking from the water what the animals give to it, *viz.*, carbonic acid gas.

That leaves absorb carbonic acid gas, we shall learn from a simple experiment. Take a bunch of fresh green leaves of a water-plant, say Utricularia, and place it under a funnel in a vessel filled with fresh spring water. Over the mouth of the funnel place a test-tube filled with water. Then expose the apparatus to strong sunlight. After a very short time you will see bubbles arise from the leaves which are collected at the top of the test-tube, as is shown in the figure. When all the water in the test-tube has been displaced, we shall close it with the thumb, take it out and introduce a glowing chip, which will at once burn very actively: the tube contained oxygen gas. This gas was evolved by the leaves under the water. They absorbed the carbonic acid gas, dissolved in fresh spring water, and retained only the carbon of it, setting free the oxygen. That this explanation is correct, *i. e.,* that the oxygen set free is derived from the decomposition of the carbonic acid gas and not from the water, may be proved by continuing our experiment. The evolution of oxygen gas will become less and less until it ceases altogether. This can only be due to the exhaustion of carbonic acid in the water.

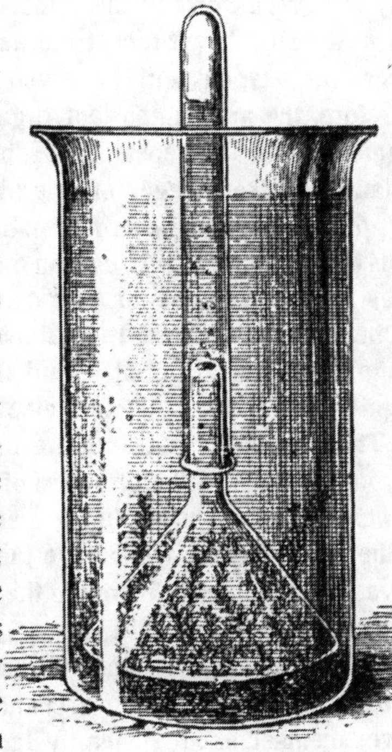

Fig. 197.—Absorption of carbonic acid gas and setting free of oxygen by a water-plant.

We know now that *plants possess the power of absorbing carbonic acid gas through their leaves and of deriving from this gas the carbon which they require for the construction of their body.*

(b) **Only the Green Parts of Plants absorb Carbonic Acid.**—It must, however, be borne in mind that not all plants

have the power of thus feeding on the carbonic acid gas in the air. It is only the *green plants* (name plants that are not green!), and only the *green parts* of these that can feed on the carbonic acid gas in the air. If we repeat the above experiment with a potato tuber, no oxygen will be given off. The green leaves are, therefore, the most important organs for nourishing the plants. Plants which are repeatedly robbed of their leaves, as for instance, by the ravages of caterpillars, become sickly and die.

(c) **The Presence of Sunshine is required.**—The green parts of plants can absorb carbon only under certain circumstances. *They require sunshine for their action.* If we place the apparatus of the experiment, described above, in a dark place, there will be no formation of bubbles, and there can, therefore, be no absorption of carbon. At night also this process cannot take place.

The fact that plants require light for their life, explains numerous features of the structure of plants: The green part of the plants are placed in the light; stems and branches, the supports of the green leaves, rise above the ground; climbers bring their leaves from the shade below to the life-giving light above; many jungle plants that would not get sufficient light on the dark-shaded ground have assumed the habit of perching on the branches of trees where they have a chance of getting more light; the leaves themselves are generally dark-green on their upper surface and whitish on the lower one; the insertion of the leaves in the stem is always such that all of them get their due share of light; those placed at the base of a stem are in many cases larger, long-petioled and flatly exposed to the light, those above, small and pressed towards the stem (Mustard, Ladies' finger); if the stem is weak and straggling (Cucumber, page 74), the petioles by twisting and bending themselves, assume such a position as to place every leaf in the light; and large leaves are often divided into smaller parts so as to let the light pass through their holes to any leaves that grow below them.

(d) **The Inner Structure of the Leaf: Chlorophyll and Photosynthesis.**—In order to understand better the manner in which leaves absorb their food, we must examine the *inner structure* of them. The illustration on p. 217 shows the vertical

section of a leaf as seen in a microscope. The upper and the under surface are formed by flat cells with thick walls. This is the outer skin or *epidermis (a)*. Between the two skins there is a layer, more or less thick, of soft and green tissue, the upper part of which consists of oblong cells, arranged at right angles to the surface, and placed so evenly parallel to each other that they have been compared to the pales of a fence. They are called *palisade*

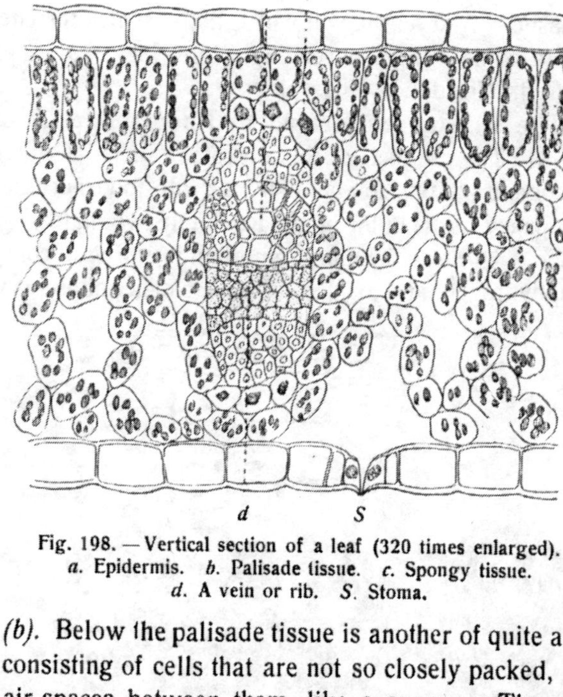

Fig. 198.—Vertical section of a leaf (320 times enlarged).
a. Epidermis. *b*. Palisade tissue. *c*. Spongy tissue.
d. A vein or rib. *S*. Stoma.

tissue (b). Below the palisade tissue is another of quite a different form, consisting of cells that are not so closely packed, but have large air-spaces between them, like a sponge. They form the *spongy tissue (c)*. The illustration also shows a bundle of other cells in the middle *(d)*. These constitute a vein running through, and supporting, the blade of the leaf. In the cells of the palisade and spongy tissues we can see a number of small, green spots. These denote the *chlorophyll-granules* that give the leaves their green colour. The palisade tissue contains a much greater quantity of them, and this is the reason why leaves are generally dark-green on the upper, and light-green on the under surface. It is in these green granules that the all-important work is done of decomposing the carbonic acid gas into its constituents, carbon and oxygen, and of forming new substances, which are used as food by plants in the same manner as animals use the materials they eat. This process is called *photosynthesis* (or carbon-*assimilation*). To allow the air to reach the inner parts of the leaf, there are the *stomata (S.)* on the under surface of the leaf, which lead into the spaces between the cells of the spongy tissue. And as the walls of cells, in which the chemical changes are going on, are exceedingly thin, they require protection. This is afforded by the epidermis.

(e) **The Substances formed by Photosynthesis or Assimilation.**—The products of this process, that is, of converting the raw food from the soil and the air into food necessary for building up new tissue in the plant, are chiefly two, namely, starch and protein. Of these the more important is *starch,* a compound of carbon, hydrogen, and oxygen. This substance abounds within

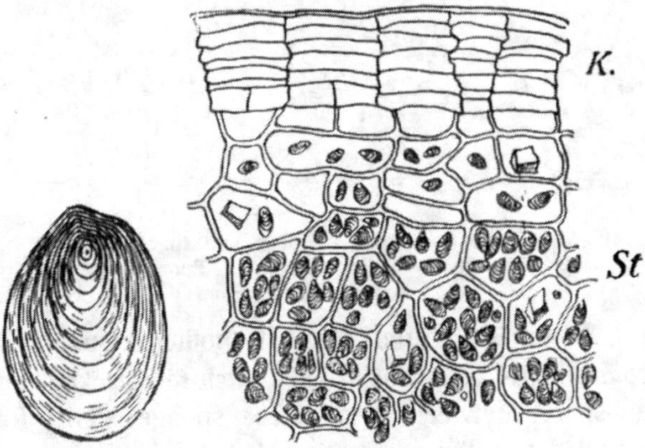

Fig. 199.—Starch in Potato.

the cells of many parts of various plants in the shape of distinct grains, as in the potato and all cereal grains, and is also the principal constituent of arrow-root and sago. *Protein* is necessary for the formation of life-substance of the plant, the protoplasm (see page 199), and is a compound of nitrogen. Other organic products formed in the cells, the little laboratories of the plants, are *grape-sugar,* found abundantly in all sweet fruits, *cane-sugar,* found in Sugar-cane, *cellulose,* a carbohydrate like starch and sugar, of which the cell-walls are built, *oils* and *fats,* common in seeds (Castor-oil, Gingili), *alkaloids* with either medicinal or poisonous, or stimulating properties (Coffee, Tea), and *acids* (Citron and many fruits).

The products prepared in the cells of the leaves, are then carried through the ribs and the stalk of the leaf and through the stem to wherever their presence is required. They are also stored up in the stem, in tubers, in bulbs, in roots, and especially in the seeds, as a reserve of material for future use.

(f) **Importance of Photosynthesis in the Household of Nature.**—Considering that by the absorption by plants of carbon dioxide exhaled by animals the balance of the constituent parts of the atmospheric air is maintained, and still more that by this process food is prepared not only for the growth of plants, but also for the maintenance of all animal and human life on earth, we understand of what immense importance the process of photosynthesis is in the household of Nature. We can safely say that without carbon-assimilation by plants there would be no organic life, and without the light of the sun there would be no photosynthesis and, therefore, no life on earth.

3. Respiration.

(a) **Proof of the Fact that Plants absorb Oxygen and set free Carbon Dioxide.**—Plants, like animals, absorb oxygen and give off carbonic acid gas, which process is known as respiration. This cannot usually be observed at the time when the process of assimilation takes place. It is, however, very evident in parts that are not green and in all parts at night. Take, for instance, two narrow-necked glass bottles of equal size and fill one of them one-third with germinating seeds of Bengal gram or flower-buds. After about a day insert a lighted taper. In the empty bottle the taper will go out after it has burnt a little while, *i. e.,* until the oxygen contained in the bottle is used up by the burning taper. In the other one the taper goes out at once,

showing that there is no oxygen in the bottle. The latter must have been absorbed by the growing seeds which gave off carbonic acid instead.

(b) **Necessity of Respiration.**—Living plants have to do work of various descriptions: they lift their parts against gravitation, they force their roots through hard soil and often break up even masonry work, they grow and form new cells, they move tendrils and press winding stems against their supports. Work always needs a supply of energy, just as in animals or in an engine. The source of power underlying the work of animals is derived from their food, the use of which is connected with their respiration by which carbon is burnt and heat is produced. So it is with plants also, though their work may be slow and easy, and hence the need of such energy less than in animals.

In the green parts the action of respiration at day cannot be shown as they assimilate under the influence of sunlight more vigorously than they respire. They rather appear to exhale oxygen. We have, therefore, to *distinguish* strictly *between Assimilation and Respiration. While in the process of assimilation green plants alone, and only in the light, decompose carbonic acid and give off oxygen,* **all** *plant-organs, without exception, both by day and by night take up oxygen and give off carbonic acid.*

When the action of respiration is vigorous, *e. g.*, in young leaves, or in wounded ones, it is indicated by a red hue (see Mango, Cinnamon), due to a red colouring matter, called erythrophyll *. This pigment is also present in flowers and fruits coloured red or blue, according as it is dissolved in an acid or alkaline cell-sap. It appears, too, in old leaves, painting the foliage of northern forests with that charming red before they drop their leaves.

If plants or parts of them are deprived of oxygen and the action of respiration is thus checked, they are hindered in their growth, become sickly, or perish. This can often be noticed in pot-plants or fruit-trees which are planted too deeply and covered with too much earth. Their roots cannot get the required air for respiration. Conversely, the loosening of the upper crust of soil is advantageous to crops to allow the air free access to the roots.

* From Greek *erythros*, red, and *phyllon*, a leaf.

(c) **Ways for Respiration.**—The entrance of oxygen into the plant-body is accomplished in the same way as that of carbonic acid for assimilation; it enters through the *stomata* of the leaves into the air-chambers inside, and is distributed in the tissues in all directions, penetrating into the protoplasm of the inner cells. Stems and stalks that are covered with bark also have their openings, called *cortical pores,* by which the free passage of gases is secured. They appear as small, brown outgrowths scattered over the surface of stems (fig. 201, see also stem of Shoeflower) and consist of loose cells with large intercellular spaces communicating with groups of cells (called medullary rays) that run through the woody tissue to the centre of stems.

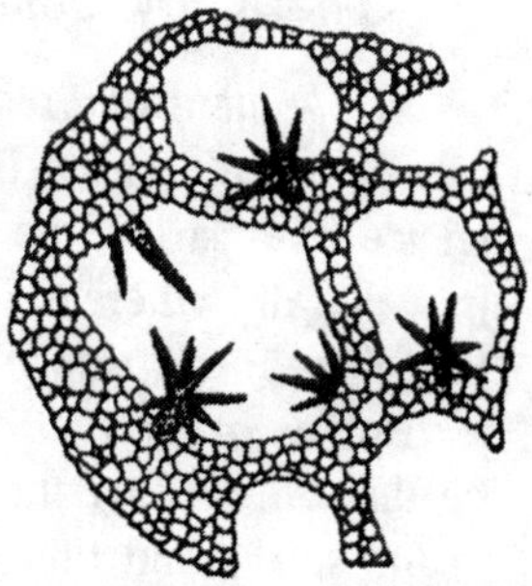

Fig. 200.—Large air-chambers in the leaf-stalk of the Water-Lily, forming air-canals from the leaf-blades to the root.

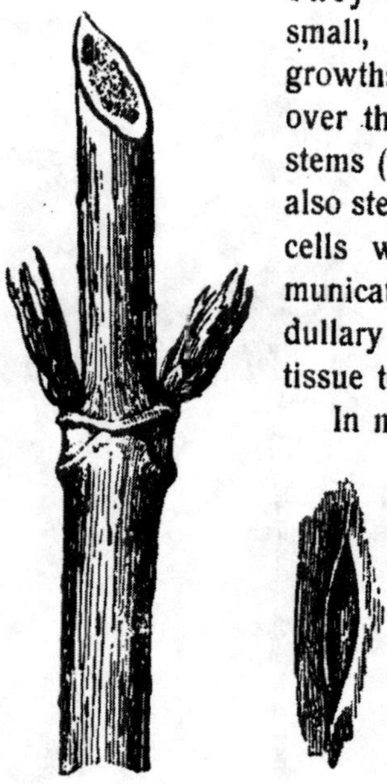

Fig. 201.—Cortical pores in a young stem of Elder. (Nat. size.) To the right a magnified pore.

In marsh and water plants, which stand partially in the air, *e. g.*, the Rice-plant and the Water-Lily, intercellular air-spaces are extensively developed and form connecting canals (compare the leaf-stalks of the Water-Lily) through which the atmospheric oxygen can reach the organs growing deep in the swampy soil which are cut off from any communication with the atmosphere. The roots of Mangrove tree obtain air to breathe by means of special growths thrown up above the level of the water (see page 56).

2.—THE ROOT

A. The Work done by the Root and its Main Types.

1. A plant derives its raw food partly from the air (Carbon dioxide) and partly from the soil (water and mineral salts). Hence one part of its body, the shoot, rises above the ground into the air, whereas its other part, the root, grows downward into the soil to *absorb water* and *mineral salts.*

But in order that the over-ground parts of a plant may not be thrown over by the wind, the plant must be firmly *anchored in the soil.* And this is another function of the root.

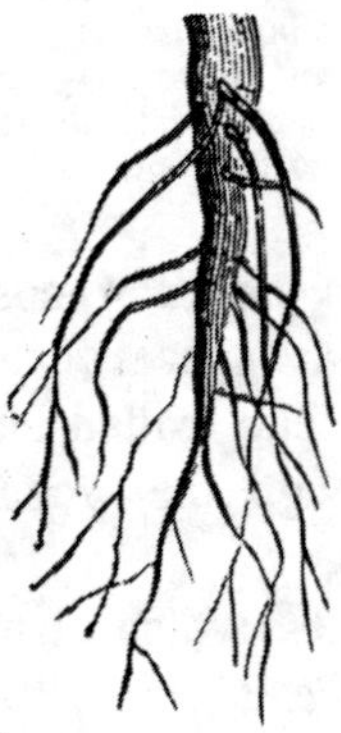

Fig. 202.—Main root with side-roots of Thorn-apple.

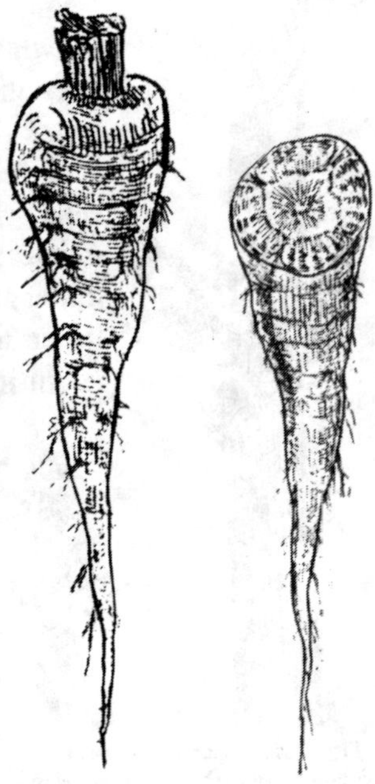

Fig. 203.—Carrot. Tap-root of Carrot with small side-roots.

2. As we can see in the germinating seed of a horse gram (p. 31) the root that forms the continuation of the hypocotyl, the *main* or *tap-root,* works its way vertically into the soil. Branching from it there issue *side-roots* in all directions, more or less slanting or even horizontally. It may be noticed, *e. g.,* in the swollen root of the Carrot, that such side-roots are arranged in vertical rows, and that the epidermis of the main root is rent where the side-roots issue. As the side-roots grow and divide into finer and finer roots, the whole soil within the reach of the plant is traversed by hundreds and thousands of rootlets.

3. In grasses and other monocotyledons the main root soon perishes, and then *adventitious roots*, produced from the lowest or any one of the lower nodes of the stem, take up the functions of the main root and its side-roots. Such adventitious roots may be formed from other parts of the plant also, *e.g.*, from underground stems (Canna, Lotus, Potato), from stolons (Hydrocotyle), from creeping stems (Ipomæa batatas), from branches (Banyan), from cut stems (Lantana, Rose), or in certain plants even from leaves (Bryophyllum, Begonia).

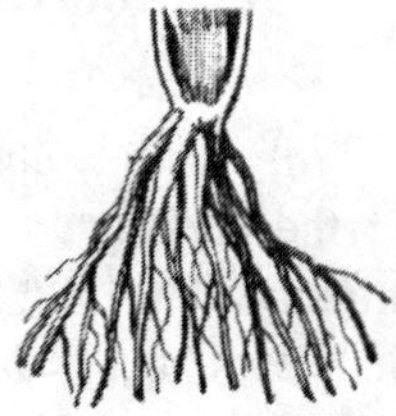

Fig. 204.—Fibrous roots.

Fig. 205.—Parasitical root of Mistletoe.

4. Some types of roots do not penetrate into the soil. As *clinging roots* they fasten the plant to its support (Pepper, Pothos scandens); as *aërial roots* they absorb the atmospheric moisture (several epiphytical orchids); as *parasitical roots* they sink their suckers into the tissue of the foster plant; and as *breathing roots* they help the Mangrove trees to get oxygen for their roots in the marshy soil in which they cannot breathe.

5. In some plants roots become *storing places* for food material and are then swollen. So the thickened tap-root of the Carrot and of the Radish, or the swollen adventitious roots of Asparagus, of the Sweet Potato, and the bulbous roots of some terrestrial orchids (Habenaria rotundifolia, p. 159).

Fig. 206.—Thickened side-roots of Asparagus.

B. The Work of the Root and its Detailed Structure.

1. Roots as a Means of keeping the upper Structure in Position.

(a) Find some healthy plants of about the same size, and cut the *main roots* of them about half an inch below the surface without disturbing the side-roots. It will be necessary to support the stems now by tying them to a stick thrust into the ground beside them. This shows that the main root is best suited to anchor the plant firmly. Trees having shallow, spreading roots and no distinct tap-roots are easier uprooted than those with a straight deep root. In proportion as the upper structure of a plant grows, it is exposed to the wind, and the more a plant is

Fig. 207.—Plank-roots of the Goldmohur tree *(Poinciana regia).*

exposed to the force of the wind, the firmer it must be rooted in the soil.

(b) But *side-roots* have their uses as strengthening supports as well, and may be compared to the ropes of a flagstaff that have to keep the flag pole in position. The way in which huge

trees are anchored in the ground by side-roots can be seen in the large buttresses at the bases of some of them. If these are cut or sawn across, it will appear that their shape is almost like planks set edgewise. "The astonishing strength of tree-roots can be imagined when we watch a tree in full leaf during a storm. As the terrific force of the gale sways the trunk to and fro, the roots are subjected to an enormous pull. Like great India rubber cables they give and retract, and when the wind subsides, we find the trunk as firm as ever."

(c) In which part the *growth* of the root takes place can be experimentally seen in the following way: we take a few germinating seeds of the Bean and mark the root from its tip with Indian ink at intervals of exactly one-eighth of an inch. Then we pin the seed to the lower side of a cork fitting into a wide-necked flask. To afford the seed the required moisture there should be a little water in the flask. After twenty-four hours we find the root to have grown considerably, and from the distances of the marks we distinctly gather that growth took place only at a short stretch above the root-end. The tip of the root is, so to say, forced into the soil.

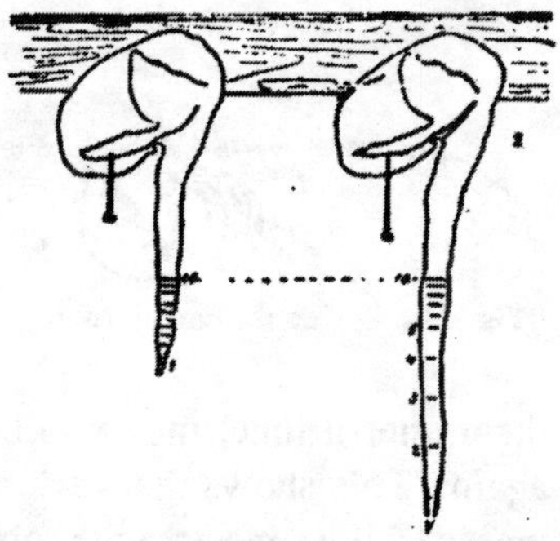

Fig. 208. — Growth of root taking place behind the root-end.

Fig. 209. — Root-cap of Pandanus.

(d) The root-tip is, however, exceedingly tender, and when it is forced into the soil so energetically, the tip must be protected from injury. Examine, with a hand-lens, the root-tip of a Tradescantia grown in water, or of a germinating Sunflower. You will find that it is covered over by a protective cap, the *root-cap.* It consists of a cellular tissue, which is constantly renewed from within to replace the old cells which get worn out and swell up jelly-like, giving the tip a slippery touch, as can be noticed in the pendulous aërial

roots of the Banyan tree. Huge root-caps of a series of scarious layers will be seen on the ends of the thick adventitious roots of Pandanus.

(e) Growth of *Tap-root* influenced by *Gravity*. – We have already seen that the tap-root grows vertically down into the ground and that this is of great advantage to the plant. In order to find whether the tap-root always grows downward, we shall make an experiment: we take a few soaked seeds in which the roots have already appeared, and put them in various positions on moist soil covering them with a tumbler. After twenty-four hours, we find the root-tips changed in their direction and turned downwards. If we now disturb them so that some point upwards, others sideways, and so on, and examine them after a time, they will be found to have turned downward again. This shows that each main root persists in growing downwards. If it meets with obstacles in its way, it resumes its original course immediately after passing them.

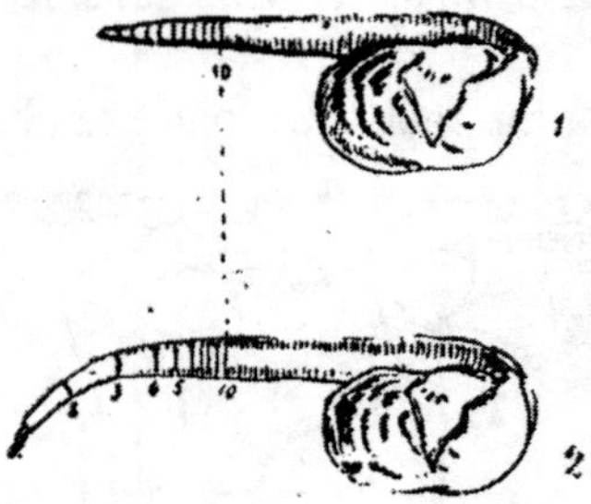

Fig. 210. — Geotropism of root.

By experiment it has been found that the downward direction of the main root is determined more by gravity than by any other influence, as light and moisture. We therefore say, the *main root* is *geotropic*.

In the same measure as the geotropism of the main root is an advantage to the plant, it is good that side-roots grow in a slanting or horizontal direction and are influenced by other forces (sensitiveness to moisture).

2. Roots as a means of Absorption.

The main function of the roots is the absorption of water and liquid food. In order to understand how roots absorb water from the soil, we must first know what the soil is and how water exists in the soil.

(a) The *Soil,* and *how Water exists in the soil.*—When a well is dug or the foundation of a house is laid, we can see that the soil has *different layers.* The topmost layer is mostly a thin-grained layer of earth, more or less black, containing humus, *i.e.,* decayed organic matter. Then follows the sub-soil with less humus and larger pebbles or pieces of rock, and beneath that there is solid rock; it may be laterite, as on the West Coast of India; or Granite and Gneiss, as in the Ghats; or vulcanic rocks, as in the Deccan; or any other of the various kinds of rock. By the mechanical action of water and wind; the changes of temperature; the chemical action of air and water; and the action of plant-roots and burrowing animals the rock, where it lies bare and open, is broken up and weathered, and soil is produced.

The *parts* of which *surface soil* exists can be found by the following experiment. Put some garden soil into a small basin with water, and stir it well. The lighter substances now come to the surface of the water: they are chiefly vegetable mould and other decayed organic matter (humus). Take the humus off. Now pour the remaining muddy water into another basin leaving the heavy and hard particles in the first basin, and allow the water in the second basin to evaporate: a crust of clay remains. Wash the remainder in the first basin again and again: the hard particles will be found to consist mainly of sand. The main parts then of which the surface soil consists are vegetable mould, clay and sand.

Now, *how does water exist in the soil?* Take three flower-pots. Fill the first with pure sand, the second with pure clay, and the third with good garden soil. Pour water in each of the pots. Which of the three retains most water? The pot with clay does not allow the water to percolate; the one filled with sand absorbs the water quickly and after a short time the water trickles through and runs out from the hole at the bottom of the pot. But in the third pot, containing a mixture of sand, clay and humus, water soaks well and is retained best. According to the proportion of sand and clay mixed in garden soil, the degree of its power of retaining water varies.

Of great importance to the fertility of the soil is the question, *how long is the soil able to retain its moisture.* Moisture is evaporated from the surface. But the rate of evaporation is very small when the surface is loose or covered by fallen leaves or hay. Another part of the water sinks deeper and deeper, and, in a certain depth, forms what may be called ground-water, when a layer is reached that does not allow the water to percolate.

(b) Exploration of Soil by Roots for Moisture.—The fact that roots follow up and grow towards moist places, is generally known. This is strikingly illustrated by keeping a flower-pot watered daily in the neighbourhood of a tree. The roots of the tree will be found to penetrate the hole at the bottom of the pot and invade the pot. We thus see that *side-roots* are *sensitive for moisture,* or in other words, that they grow stronger towards moist parts of the soil than in the direction of dry soil.

There is, however, a great difference in the way in which various plants explore the soil for moisture. Some grow their tap-roots right down and fetch their water from the deep; in other plants the tap-root does not grow vigorously, and their side-roots keep near the surface, feeding on the surface moisture. The former may be called *deep-ground-feeders,* and the latter *surface-feeders.* The difference between deep-ground-feeders and surface-feeders is noticed in a coffee-plantation, where coffee-plants, which are surface-feeders, are not harmed in any way by the presence of the large shade trees in their vicinity, as these are deep-ground feeders and do not interfere with the supply of moisture to the coffee shrubs. Casuarina trees, to mention another example, are surface-feeders, and hence the growth of annuals near such trees is greatly hampered.

The extent to which the *roots* of many plants *spread* in the soil is beautifully *suited* to the *way* in which the *rain-water* is *conducted down* to the soil by the leaves. In some cases, as in the Carrot and other plants with swollen roots, the leaves conduct the rain water towards the centre of the plant, where it runs down the stem to the tap-root. Such tap-roots have only small side-roots. In other plants (see Mango tree, page 27) the water is conducted from leaf to leaf towards the periphery of the tree and there falls to the ground; and in these plants the ends of the side-roots will often be found just beneath the periphery of their crowns. In the same measure as the crown of the tree increases in width, its roots grow to have the full benefit of the water falling from the leaves.

(c) Root-hairs.—Allow some seeds of maize or mustard to germinate between sheets of moist blotting paper. Put each plant into a phial filled with water fixing its stem in the neck of the phial with a cotton stopper. Examine the roots and you

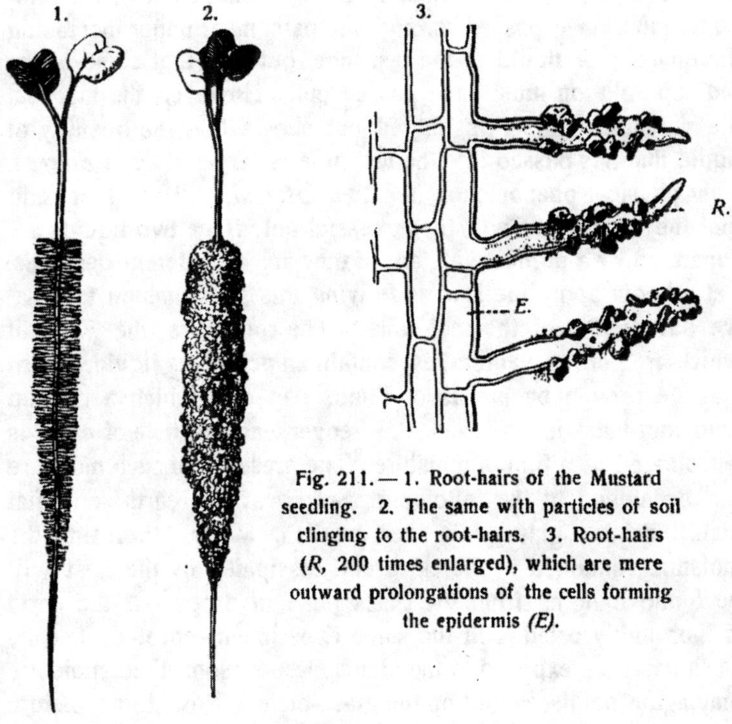

Fig. 211.—1. Root-hairs of the Mustard seedling. 2. The same with particles of soil clinging to the root-hairs. 3. Root-hairs (*R*, 200 times enlarged), which are mere outward prolongations of the cells forming the epidermis *(E)*.

will find a delicate down just a little above the end of each tap-root and of each side root. This down consists of numerous little hairs or threads. When a seedling is carefully taken out of the soil, the root-hairs cannot be distinctly seen as they are too thickly covered with earth (see fig. 211, 2). Even if washed in water, the root-hairs cannot be freed from the particles of earth. So closely do they cling to them,—you would sooner break off the hairs than separate the mud particles from the hairs. Now what may the function of these root-hairs be?

Take a test-tube and fill it with a saturated lotion of salt or sugar, stained red by eosin. Close its mouth by means of moist parchment paper, taking care that no air bubbles are left in the test-tube. This apparatus is now immersed in a dish with rain water. After about twenty-four hours it will be observed that the parchment paper is bulged out considerably and that the rain-water is red, thus indicating not only that some of the rain-water must have passed through the parchment paper increasing the quantity of liquid in the test-tube, but also that some of the red salt solution must have passed out. However, the fact that the parchment paper is bulged out shows that the quantity of liquid that has passed into the test-tube is larger than *vice versa*. This physical phenomenon is called *Osmosis*. It must be said that the experiment will be successful only if the two liquids are separated by a porous wall, and if they are of different densities. Let us now apply the law underlying this phenomenon to what we have learnt of the root-hairs. The root-hairs, the walls of which are porous membranes, contain an acidulous liquid, which may be proved by means of litmus paper in which a root-tip with root-hairs is squeezed. Moreover, each particle of earth is surrounded by a film of moisture. The presence of such moisture can be shown in the following way: heat dry earth in a flat dish, placing a glass plate over the dish. After a short time the moisture contained in the earth and dissipated by the heat will be found hanging from the glass plate in drops. If the earth is thoroughly dried, and the same experiment repeated after the earth has been exposed to the atmosphere for some time, moisture may again be discovered on the glass-plate. Now, this moisture passes through the walls of the root-hairs, and a small quantity of the acid inside the cells passes out and helps to dissolve the minerals of the earth. The quantity of moisture absorbed by the root-hairs being greater than the quantity of liquid secreted by them, the root-hairs become turgid. In the same way the liquid food is passed on from cell to cell in the body of the plant, the cells high up being filled with a denser cell-liquid than those below, owing to the action of transpiration. If there is a deficiency of moisture in the soil, and the loss of moisture in the cells of the upper structure cannot be made good by the rising flow of liquid, the turgor of these cells will be diminished. The cells cease to be stiff and become limp, and the whole plant, or at least those parts which are not stiffened by woody cell-walls, will fade.

A plant will also fade, if it is transplanted and care is not taken to leave the root-ends undisturbed. This is explained by the fact that the plant deprived of its root-hairs cannot absorb moisture from the soil.

(d) Manuring and *Rotation of Crops.*—As we have seen on page 213, the minerals absorbed by the roots are nitrogen, sulphur, phosphorus, kalium, natrium, calcium, silicium, magnesium and iron. All plants do not feed on exactly the same elements. Moreover, some require more of one, and others more of another element. But all require a large quantity of nitrogen, which they derive from various nitrogenous salts. Now, farmers who year after year grow plants on the same fields, must take pains to *manure* their fields, if their crops shall not deteriorate. For, with the crop they remove a great quantity of raw plant-food from the soil, so that the soil becomes poorer and poorer in the materials available for plant-growth. The manure used ordinarily is decaying or decayed animal and vegetable substances. Since the various requirements of different crops have been investigated scientifically, manuring can be economised by the use of artificial manure and richer crops can be secured thereby.

As some plants require more or less of a particular element than others, and strike their roots more or less deep in the ground, the farmer can *rotate his crops,* that is to say, if paddy and pulses, for instance, are grown in a certain field one year, they are followed up the next year by sugar-cane, and so on, coming back again to paddy eventually.

3. THE STEM

The work of the stem can be said to be threefold. It has

1. to support the branches and leaves, and to spread them out to the air and the sun;

2. to carry the sap from the roots up to all parts of the plant, and to bring the elaborate food, formed in the green parts of the plant, down to the points where growth takes place;

3. to serve as a food-store for use of the plant in the future.

A. Stems as Means to spread out the Leaves and Flowers to the Air and the Sun.

We have learned that the inner cells of leaves are workshops wherein the plant prepares materials for the construction of its body. As this can only be done under the influence of sunlight, and also as carbon can only be taken from the air, it follows that the leaves must be freely spread out to the sunlight and to the air. The same is necessary for the flowers also in order that they may be pollinated by the agency of insects or the wind so as to produce fruit. Stems, therefore, rise up, and, in many cases, form branches on which they support a great number of leaves and flowers.

I. Growth and Description of Stems.

The topmost end of the stem consists of very tender cells which, by division, multiply at a quick rate. Stems, therefore, grow there in length, and we call these points *points of growth.* As the stem grows, leaves are developed on its sides. The places where the leaves rise are often swollen and are, therefore, denoted *nodes,* the parts between the nodes being called *internodes.* The internodes are of different lengths in different plants: they are generally long in grasses, but short in plants forming leaf-rosettes (see Elephantopus, p. 206).

The internode may be filiform = thread-like (Horse Gram); or succulent = fleshy (Balsam); or nodose = swollen at the nodes (Adhatoda); or tuberose = swollen like a tuber (Knolkohl); or articulate = jointed (Casuarina); or winged (Desmodium triquetrum). Its section is round in Maize, square in the Labiatæ, triangular in the Sedges, and oval in Ragi.

2. Buds.

The growing point of the stem is folded over by young leaves, the inner ones being more or less covered by the outer ones and thus all forming a bud. The bud is, therefore, *an undeveloped shoot*. If buds are to rest for a time, as the winter buds in cold climates and the summer buds in tropical countries, they are often specially protected by hairy scales. Such buds are termed *closed buds,* in contradistinction to naked buds without such protective scales. (Resting buds with scales may be looked for in Diospyros embryopteris, *Kan*. Bandha.)

In *naked buds* the tender parts of the growing point are sometimes protected by stipules (Rose, Banyan), or by the leaf-bases of old leaves (Garcinia morella, *Kan*. Jārige), but in most cases by the blades of older leaves, folded over the younger leaves in various ways (see *æstivation* of floral leaves).

Moreover, each leaf is folded or packed in its bud *(vernation)* in a way peculiar to the plant. In the Rose and in Argyreia, they are folded over their mid-rib like a sheet of note-paper; in the Palmyra Palm they are folded over their several basal veins like a fan; their blades are rolled from one side to the other in Musa or Canna, from both edges inwards towards the mid-rib in Lotus, or backwards, in Jujube, and from top to bottom in the Ferns.

Besides the terminal buds at the ends of the stems and their branches, there are *axillary buds,* borne in the axils of leaves, from which branches originate eventually. Axillary buds do not all open, but sometimes remain dormant or sleeping, until one day the top of the stem is cut off. Some trees, as Palms, never develop any other than terminal buds.

Young shoots are often specially protected from intense light and heat by their hanging down quite limp (Mango, page 26) or by placing their leaves vertically up (Banyan, Jack, page 121), or by a red hue in their tender leaves (Cinnamon, page 129).

3. Direction of Growth influenced by Gravity and Light.

(a) By gravity.—Plant a few germinating seeds of Horse Gram in various positions, some with their hypocotyl up, others with their hypocotyl down, and others again with their hypocotyl horizontal: the stems of all these plants will invariably rise vertically up when they grow. Observe a cocoanut tree thrown over by a storm: its shoot will raise its head straight up, after a time. Plant cuttings of Croton in a slanting position: the new shoots will grow upright. Stems, as a rule, grow erect, even on sloping ground. This is due to gravity. Gravity causes such stems that are disturbed in their natural vertical growth to grow stronger on their lower side till they regain their original direction of growth.

The main root grows in the opposite direction of the stem, following the law of gravity similarly. The main root is, therefore, said to be geotropic, and stems are *negatively geotropic.*

The branches of a tree cannot grow straight up likewise; owing mainly to the influence of light, they grow obliquely or even horizontally.

(b) Light.—Stems planted near a house do not grow straight up, but incline away from the house towards the light. Light causes the shaded part of a shoot to lengthen, more than the part exposed to it, and thus the direction of growth is changed. A plant growing free in the open air is exposed to the light all round in the same degree, and hence all parts of the stem grow equally, and the stem becomes erect.

Roots, on the other hand, show the opposite tendency. In the hanging roots of the banyan tree, for instance, it may be noticed that they bend towards the dark side, *i. e.*, away from the light. Stems are, therefore, called *heliotropic,* and roots *negatively heliotropic.*

4. Habits of Stems.

Most stems are *erect* (Sunflower, Mango). But some are weak and lie on the ground raising their heads gradually: *ascendent* (Desmodium triquetrum); or so weak as to lie flat on the ground:

procumbent or prostrate (Evolvulus alsinoides). *Creeping* stems also lie prostrate on the ground, but form roots from their nodes (Ipomœa biloba).

Other weak stems assume the habit of *climbing* other plants in the shade of which they grow, and thus raise their foliage to the life-giving light. From the various methods by which they climb, they can be classified as follows:—

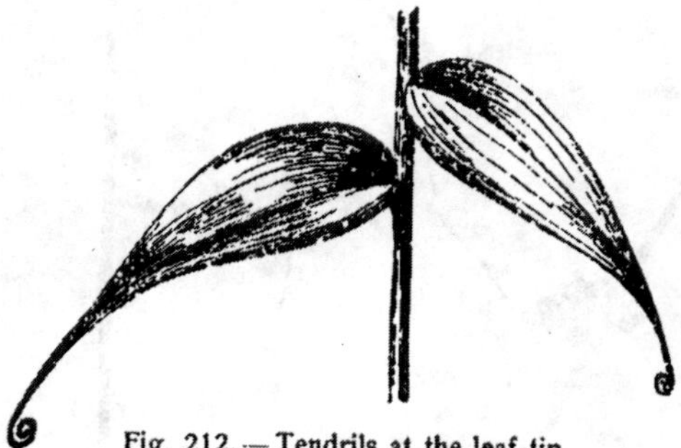

Fig. 212.—Tendrils at the leaf-tip of *Gloriosa superba*.

1. *Root-climbers, e.g.,* the Pepper vine; these produce little clinging roots on the lower side of their stems, *i. e.*, on that turned away from the light, to attach themselves to the supporting trunk of a tree or to a wall.

Fig. 213.—*Lophospermum scandens*. The leaf-stalk used as tendril.

2. *Tendril-climbers, e.g.*, the Pea, the Cucumber, the Gloriosa. These attach themselves to their supports by means of tendrils, which are sometimes merely the tapering ends of the mid-rib (Pea, Gloriosa), sometimes the leaf-stalks (Potato creeper and fig. 213), sometimes branches (Grape vine; Hippocratea Grahamii; *Kan*. Padiri), and sometimes separate organs growing from the axils of the leaves (Cucumber).

3. *Twining Climbers.*—The stems of these climbers wind round their supports, some in the direction in which the hands of a watch move (Yam), and others in the opposite direction (Bindweeds, Shankapushpa). If unwound and turned the other way, the young parts will insist on following their original direction.

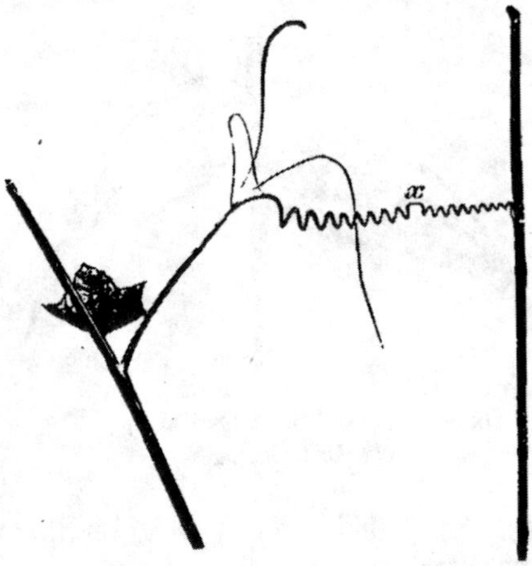

Fig. 214.—Tendril of *Luffa acutangula.*

4. *Scramblers, e. g.,* the Rose, the Wild Asparagus, Lantana, Bougainvillea, etc. These climbers stretch their tips through the holes of the thicket in which they grow, and then open their branches wide so that they may not glide down. Hooks and spines are often made use of to the same purpose.

A common feature of these climbers is their quick growth and the absence of leaves and branches at their tips, called forerunning tips (see page 87).

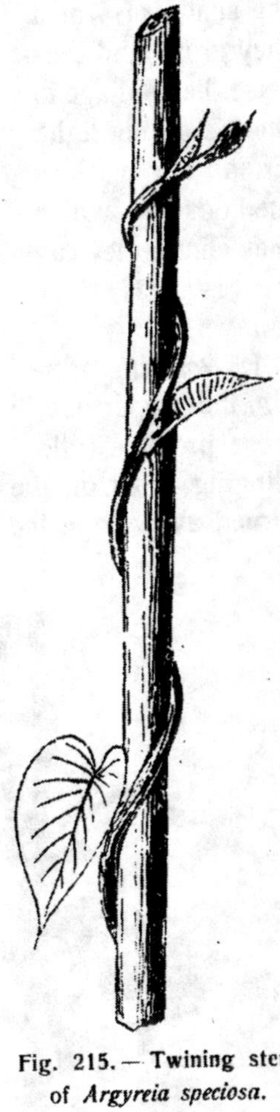

Fig. 215.—Twining stem of *Argyreia speciosa.*

5. Strength of Stems.

As the stem with its heavy load of leaves, flowers and fruits, high up in the air, is subjected to the bending, pulling and shearing stress of wind and storm, it must have a certain amount of strength to withstand such stress or possess some other means to lessen the destructive effect of the wind.

Stems are ordinarily not strong enough to stand against hurricanes; their leaves are torn away, branches broken, yea, whole trees uprooted and destroyed. The destructive effect of such a tempest can be specially seen on isolated trees. If they

Fig. 216 — Trees deformed by the effect of the sea-breeze.

stand in groups, they help one another and resist the storm-wind with their united strength. Winds also change the mode of the growth of trees and imprint in their very faces signs of the hard battle which they have continually to endure. Such trees follow with their branches the direction of the prevailing winds (fig. 216), or spread out their branches in horizontal sheets instead of growing high and increasing the plane of resistance (page 45). To a certain degree every plant is enabled to protect itself against the destructive stress of the wind: by means of their petioles they place their many leaves in any direction the wind blows, and thus escape most of the wind-pressure; long linear leaves, as those of the Grass family, wave like a flag in the wind without opposing it; large leaves as those of the Cocoanut tree are slit into numerous segments to allow the wind to pass through; the Plantain allows the wind to tear its broad blade into many pieces for the same purpose. In spite of all such means there remains a great deal of pressure to which a stem is subjected, and it follows that, to bear up against this pressure, it must be specially strengthened. How much the strength of a stem is strained, can be seen in the culm of a Rice plant, which, with a diameter of five millimetres at its base, grows to a length of one thousand five hundred millimetres; it is indeed marvellous to see that this slender stem with its heavy load of leaves and grains is not broken when the wind blows at it. What, then, makes the stem of this or any other plant strong enough to hold its own against the weight of the whole plant which presses it downward, against the bending stress of the wind, which tends to break it, and against the pulling power of the wind which tends to uproot the plant?

First of all, it is the thick-walled *woody cells* which not only make up the greater part of the trunk of a tree, but are also present in the annual stems of many herbs. Of equal importance to the strength of the stem are the *bast-cells,* commonly known as fibres, which are to be found in the inner bark of the stems of dicotyledons (see Flax, page 20) as well as in the fibrous bundles of monocotyledons (p. 138). The strength with which the fibres of plants in general resist any pull is equal to that of the best wrought-iron and, in certain plants, even exceeds that of steel.

The arrangement of these cells in the stem is something that makes every student of Botany marvel at the wisdom displayed in the works of nature. The principles of architecture, which man took many centuries to discover, after many trials, are here exhibited in their simple original beauty. In the description of the culm of a Rice plant we have seen that the sides of the stem are exposed to the greatest stress and the middle portion to the least; hence the culm is hollow to save material and the sides are strengthened by a cylinder of strong fibres. Similarly the stem of the Labiatæ and many other plants have their four corners strengthened by strong cells. The strength of beams is proportional to the breadth and to the square of the depth. This rule is illustrated by the edgewise placement of leaf-ribs, *e. g.*, in the Teak, or of plank-roots (see fig. 207). If other plants are examined in this respect, we always find that, though the position of their fibres may be modified, they always answer the fundamental principles of architectural structure.

When the crown of a tree is shaken and the stem is bent by a storm, the roots have to sustain an enormous pull, just like the cables that are used to keep a vessel at anchor. If such cables were untied and their several strands were made to hold the ship, they would be easily torn one by one by the movements of the vessel. As they are united into cables, the pull which would tear the single strands is equally distributed over every one of them, and thus the cable is able to withstand it. Thus in roots we find the woody and bast-cells crowded together in the middle, which makes them strong like cables. Similar arrangements are found in the climbing stems which are also exposed to strong pulls.

6. Various Types of Stems.

The greater the load of leaves a stem has to bear, the stronger must be its structure. *Herbs* are comparatively small plants of a year's growth, or of perennial growth, if their underground parts remain alive at the time when their overground shoots perish. Such herbs have soft or herbaceous stems. Hollow stems, articulated by solid and swollen nodes, as we find them, in Grasses, are termed *culms*. If a stem consists of one long internode only, and bears no leaves but flowers only at its top, as in Crinum, it is called a *scape*. The stem that does not die at the end of a season, but lasts for years, becomes woody. If such a woody stem branches off from the ground, it is called a *shrub*

(Ixora coccinea); if it has a distinct woody trunk, scarcely branching from the base and of considerable size, we call it a *tree* (Mango). The *Palm* type has an unbranched, cylindrical, woody stem with a crown of large leaves. *Lianas,* again, are woody *climbers* (Entada scandens).

7. Modes of Branching.

The axis of a plant terminates in a bud which elongates and grows leaves on its sides. In the axil of each leaf an axillary bud is developed, which may or may not grow into a branch. This is the normal growth of a phanerogamous plant, and is called *monopodial.* In some cryptogams (*e. g.*, Selaginella) the growing point is regularly forked after a short stretch. This mode of branching is termed *dichotomous.*

Fig. 217. —The "Umbrella tree" *Poinciana regia.*

A typical example of monopodial growth is the Pine tree of cooler climates or the beautiful tree Hopea parviflora (*Kan.* Bhogi, *Tu.* Bovu) of the West Coast of India. The crown of these trees naturally becomes conical, the main axis growing stronger than the branches.

But in many of our ordinary trees the terminal bud of the primary axis develops a flower or perishes, one or more branches continuing the growth of the primary axis in a more or less slanting direction, thus giving the crown of the tree a rounded or flat shape (Mango, Poinciana). If only the branch nearest the perished terminal bud is furthered, that branch taking over the direction of the main axis, and this mode is repeated after regular intervals, the main axis is called a *sympodium*. This is seen in any species of the Vine family, where the axis terminates successively in a tendril and is continued by the axillary branch growing from the axil of the last leaf pushing the tendril aside. In Mirabilis dichotoma (the Four-o'clock plant), on the other hand, the terminal bud dies, and as the plant has opposite leaves, both the axillary buds of the last pair of leaves develop into branches. This is repeated several times, and a *false dichotomy* is formed.

Fig. 218.— A branch of the Grape Vine.

Axillary buds often remain latent or dormant, until one day the top is cut off, when they awake and begin to grow. Again, twigs which are robbed of light by other branches, starve and die, so that, as the tree grows, many twigs dry up. The leafy crown of the tree thus becomes a hollow shape supported by the bare branches like an umbrella with its ribs.

8. Metamorphosed Stems.

Stems sometimes have special functions other than those mentioned on page 232, and in such cases their structure and shape are generally suited to their special functions.

The *leaf-like swollen stems* of Opuntia (p. 57) for example, take over the function of the leaves which are absent during the greater part of the year. At the same time they form a store of water and food. Similar structures are found in several species of the Cactus and Spurge families.

As *stolons* and *runners* they are a means of vegetative propagation. These are branches with long internodes creeping along or under the ground and shooting out roots at their nodes, thus forming new stations at a distance from the mother plant. The

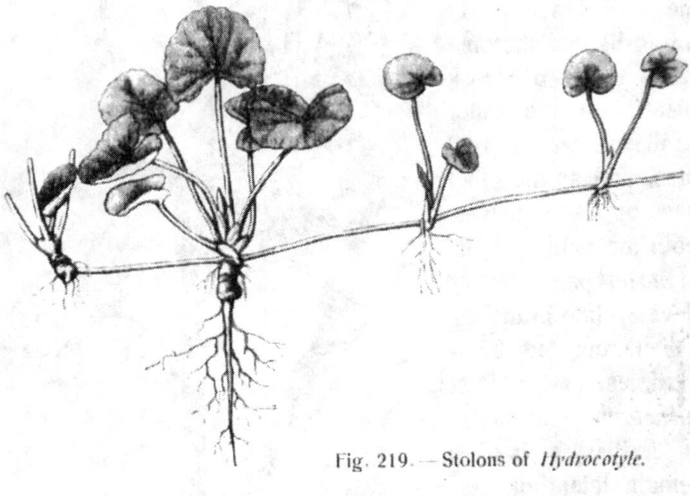

Fig. 219.—Stolons of *Hydrocotyle.*

illustration (fig. 219) shows a plant, common on the brink of paddy-fields. The strong plant in the middle is the mother plant, which has sent out horizontal runners or stolons forming young plants from their nodes. Other examples of stolons are Spinifex (p. 179) and Potato (p. 91).

In Flacourtia sepiaria, Randia, Canthium, Carissa and a number of similar shrubs found in the jungles, axillary branches are modified into *thorns* or spines, and form a means of protection from damage by herbivorous animals.

Fig. 220. — Spine of *Flacourtia sepiaria*.

In the Passion-flower branches are changed into *tendrils* for climbing. The tendrils of the vine are the ends of the main shoots (see p. 241), those of Antigonon are the stalks of the inflorescence.

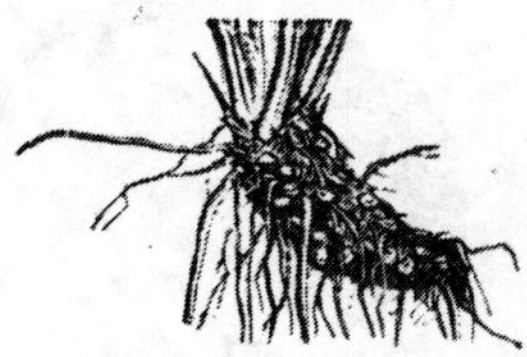

Fig. 221. Root-stock of Water-Lily.

By means of *underground shoots* perennial herbs are enabled to live during the adverse season. The three chief types of such stems are:—

(*a*) the Root-stock or Rhizome (Canna, Ginger),
(*b*) the Bulb (Onion),
(*c*) the Tuber (Potato).

The *Root-stock* or *Rhizome* (see Water-lily, p. 4; Ginger, p. 164) is either a short vertical stock with a crowded leaf-rosette as in Elephantopus, or a horizontally creeping stem of monopodial growth as in most Ferns, or one of sympodial growth as in Canna. Examine, for instance the creeping rhizome of Canna and notice the brown scarious scales covering it, thick adventitious roots growing from its lower part and breaking

Fig. 222. -- Rhizome of Ginger.

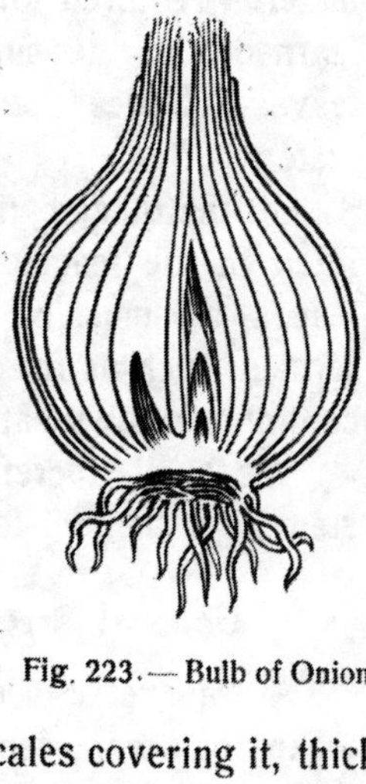

Fig. 223. — Bulb of Onion.

through the scales. Its axis suddenly turns up and gives rise to an overground shoot, whilst the creeping stem is continued in its original direction by a new branch-shoot, side-shoots also being given off occasionally.

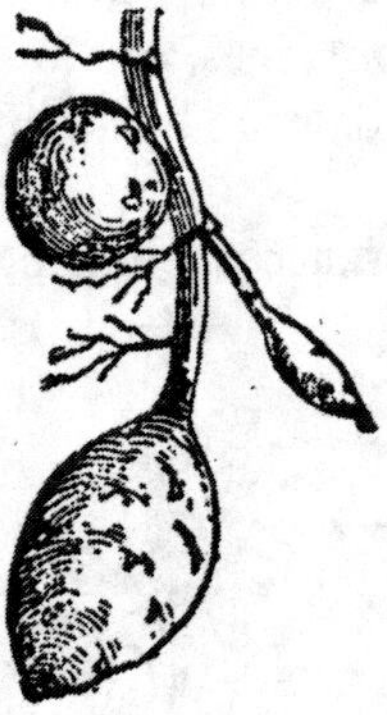

Fig. 224.
Tubers of Potato.

The *Bulb* (see Onion, p. 154) is a flat, solid stem with large succulent overlapping scales and leaves.

The *Tuber* (see Potato, p. 91) is a fleshy, roundish mass with small scales (or scars of such) and small buds in their axils. Some tubers, *e. g.*, of Taro, have one bud only at their apex.

B. Stems as Channels for the Ascending and the Descending Sap.

We have heard of the water-current from the root upwards to the leaves where the water and the mineral substances dissolved in it are used for the formation of starch and other material required for the building up of tissues. We have also learned that the superfluous water is evaporated through the leaves. Hence there must be a continuous upward flow of water.

On the other hand, we have seen that the substances elaborated in the leaves cannot remain useless in the cells of the leaves, but must be carried to those places where new tissue is formed, such as the tips of the roots. Hence, there must be a continuous downward flow of liquid food stuff.

We shall, therefore, now proceed to make a study of the stem as a channel for the ascending and descending sap.

1. General Structure and Leading Tissues of Stems.

A transverse section of a herbaceous stem, examined by means of a hand-lens, exhibits the following leading tissues: In the outer layer of cells (fig. 225, *E*) we recognize the *epidermis* which we have seen as the covering of leaves. The area surrounded by the epidermis *(C. T.)* consists of a *cellular tissue* with several marked groups of a separate tissue, known as *vascular bundles (V.)*. These bundles are variously grouped in the two

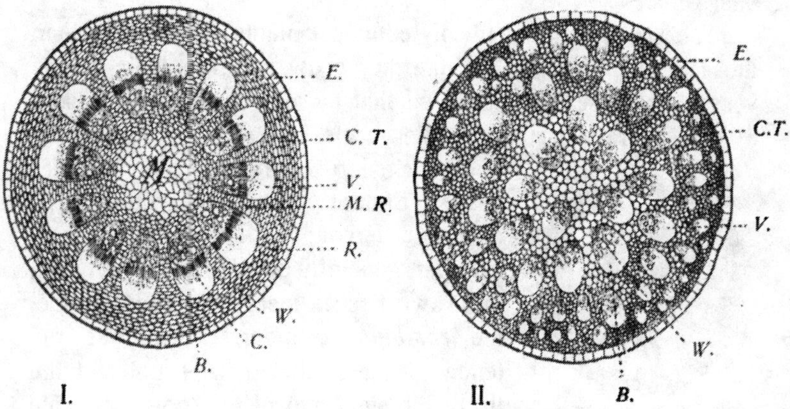

Fig. 225.—Diagrams of young stems: I. of a dicotyledonous, II. of a monocotyledonous tree. *E.* **Epidermis,** *C. T.* **Cellular tissue.** *V.* **Fibro-vascular bundles. The latter consist of an inner woody part** *(W)* **and an outer part called Bast** *(B)*. **In the dicotyledonous tree there is a Cambium layer** *(C)* **between the two parts. The Cellular tissue** *(C. T.)* **will in older trees be differentiated into Pith** *(M)*, **Medullary Rays** *(M. R.)*, **and Bark** *(R)*.

chief classes of flowering plants, *i. e.*, the dicotyledons and the monocotyledons.

(a) The arrangement of the vascular bundles in **dicotyledons** can be nicely seen in sections of the stem of Aristolochia: they are located radially in a circle. The cellular tissue is thereby divided into two distinct parts: the *pith (M)*, which lies inside, and the *rind* or cortex *(R)*, which lies outside the ring of vascular bundles. Those parts of the cellular tissue, which separate the several bundles and connect pith with cortex, are called *medullary rays (M. R.)*.

(b) On sections of **monocotyledonous** stems (Palm tree, Maize) we can see that the vascular bundles are scattered irregularly in the cellular tissue. Hence there is no distinct division of the cellular tissue into pith, cortex and medullary rays.

2. The Vascular Bundles.

These consist of two parts: an inner part made up of strong, woody *vessels,* and an outer part made up of both *soft* and *hard bast* cells.

The stems of older (dicotyledonous) plants are stronger than those of young plants. While they grow in girth, their woody vessels increase in number so that by and by the woody parts of the several vascular bundles coalesce and form one woody body enclosing some of the pithy cellular tissue. Similarly the fibrous outer parts of the different bundles unite into one cylinder, generally denoted bark. Between these two tissues there is a tender tissue, the *cambium* cylinder, which like the tender tissue in the growing point of the tip of the stem and of the root is capable of forming fresh cells by division. And, indeed, it is due to the activity of these cells that stems grow in girth: new cells are added from the cambium cylinder both towards the interior cylinder of woody vessels and towards the exterior cylinder of rind. (See also page 109.)

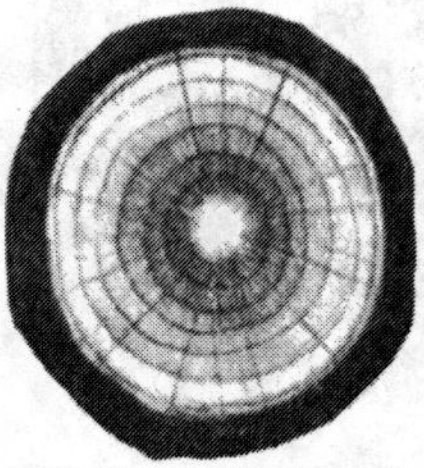

Fig. 226. — Section of the trunk of a Mango tree showing the light alburnum below the bark, the dark heart-wood in the inner layers, the pith in the centre, and medullary rays stretching from the bark into centre of the wood.

The fibro-vascular bundles of monocotyledons having no cambium, these plants do not increase in girth.

3. The Channels for the Ascending and the Descending Sap.

(a) **The Ascending Sap.** — The water-current cannot ascend in the heart-wood, the cells of which are generally impregnated with waste products, such as gum and tannin, and therefore impermeable. The dark heart-wood is thus made durable and resists the attacks of fungi which would otherwise cause the wood to decay. This property makes that part of the wood not only useful for the carpenter, but also strong enough to support

the mighty weight of the branches above the trunk. Some trees, indeed, do not contain such heart-wood, as will be seen in the White Dammar (*Vateria indica; Kan.* Dhūpa), and, therefore, become hollow when old, the dead wood being destroyed by fungi.

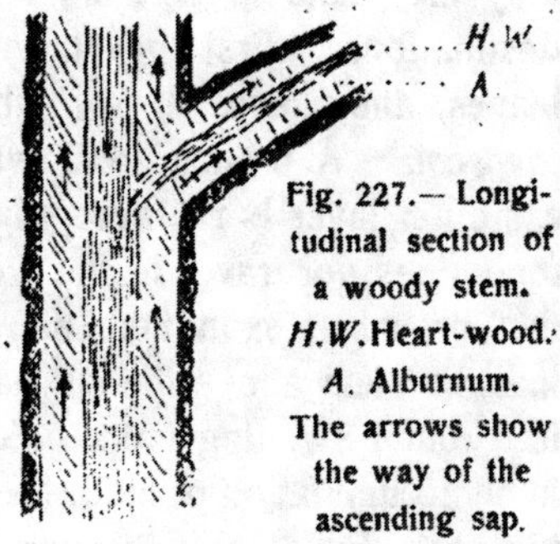

Fig. 227.— Longitudinal section of a woody stem. *H.W.* Heart-wood. *A.* Alburnum. The arrows show the way of the ascending sap.

The water-current does not rise in the bark also, as can be shown by removing a ring of bark from a branch: the leaves above the wound remain for a time as fresh and green as those below the wound. If, however, the fresh wood below the bark (alburnum) were removed and the bark allowed to stand, the leaves and all the parts above the cut would perish.

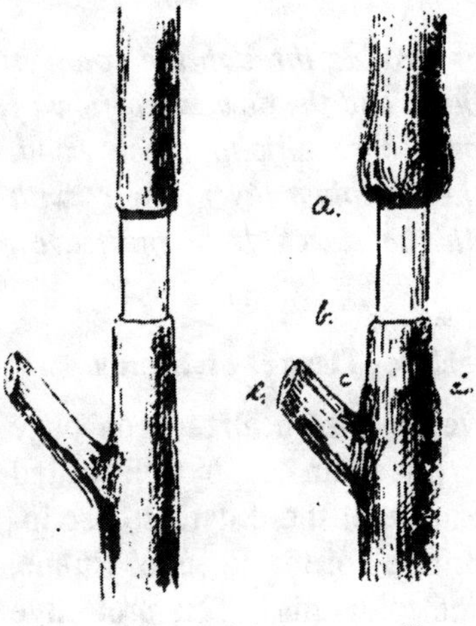

Fig. 228. — A ring of bark removed from a stem, *A.* as it appears in the first year, *B.* as it appears after 2 years. *a b* the bark stripped off, above *a* the stem is swollen by the descending sap, *b c* dry bark, below *c d* the stem has grown in size due to the sap, descending from branch *e.*

This shows that the *sap ascends* only *in the alburnum* consisting of vessels through which the sap can easily pass.

(b) **The Descending Sap.** The organic compounds formed in the leaves, of water and carbon, have to be distributed to the places where they are required, *viz.*, the growing points — the tips of stems and roots, flower-buds, fruits, seeds, the cambium cylinder, and all other places of growth. What the leaves elaborate is mostly starch. But starch cannot pass from cell to cell through the cell-walls. It is, therefore, turned into a liquid, *i. e.*, into grape-sugar, which easily goes through thin membranes like cell-walls. Besides starch, nitro-

genous compounds are formed in the leaves. These too have to be conveyed to the growing points. So the question arises which way they take to descend from the leaves to their respective destinations. First they travel from cell to cell to the veins of leaves, and thence through the leaf-stalks to the *inner bark of the stem.* A branch from which a ring of bark down to the cambium layer is removed (fig. 228) will show that the made-up food does not travel in the wood in which the sap ascends. If that branch is examined after one or two years, it will be seen that the stem above the cut has increased in size and that there is a round swelling immediately above the wound. This is due to an accumulation of food, the passage of which was arrested by the cut. Eventually the part of the wood which is laid open will probably be destroyed by parasitic organisms, so that the whole of the branch above it dies.

The upward water-current, then, travels through the younger layers of wood neighbouring the bark, and the downward flow of elaborated food travels in the inner bark adjoining the wood. Between both of them there is the cambium layer where, with the sap from the ground and with the starch from the leaves, new wood is formed.

4. The Covering or Protective Tissue of Stems.

(a) **The Epidermis.**—As we have seen already on page 245, young stems are covered over by a thin coat of cells, called epidermis, the surface of which can be of the different descriptions shown on page 204. The prickles of the Rose, of Rubus, and Lantana are also growths of the epidermis. This protective covering is sufficient for herbaceous and annual stems. When it becomes older, as in woody stems, the epidermis breaks up and falls off in thin shreds.

(b) **The Cork.**—But before the epidermis is shed, a new protective covering is prepared beneath the epidermis by the formation of cork out of the outer cells of the rind, which obtain corky walls and gradually die off. The inner cells remain active and constantly supply new cork-cells to replace those scaling off on the outside. If the layer of cork-cells is thin, the surface of

the bark becomes smooth as in the Banyan tree. In some trees, *e. g.*, the Mango, layers of cork are formed deeper inside the cortical tissue with the result that any rind tissue lying outside such cork is starved and perishes, and thus increases the thickness of the cork.

The *Outer Bark,* as the dead cork-cells together with the dead tissue of the inner bark is called, comes off it may be in flakes, as in the Jack, or in long vertical stripes, as in the Eucalyptus or Casuarina, or in horizontal shreds, as in the Rubber tree, or in rounded and hollow shell-like pieces, as in the Awla tree, or in thick rugged pieces, as in the Mango tree.

(c) **The Cortical Pores.**—The inner part of a stem requires aëration no less than a leaf. Hence the epidermis in young stems is supplied with air pores (stomata). When the epidermis gives way to a cork-coat, new ways of air communications must be opened. Such are the cortical pores or lenticels, formed by loosely arranged tissue, through which the air can pass in and out. Such cortical pores are very distinct in the stems of Alstonia as small brown or gray pustules. (See also page 221.)

Fig. 229.—Cortical pores in a young stem of Elder. (Nat. size.) To the right a magnified pore.

(d) **The Covering of Wounds.**—When leaves become old and unfit for their functions, they are dropped. But before they fall, a corky layer is formed across the leaf-stalk at its base, and when the leaf is disrupted, there remains no wound on the surface of the stem. If a stem is wounded, the living cells adjoining the wound form cork-cells to close it as soon as possible. If the wound is so deep as to lay bare the wood, the cortical tissue round the wound grows vigorously and covers the wound. This

can be seen on the surface of cut branches, on which the growth forms rounded lumps of bark on the cut surface. If the cut is somewhat distant from the axis from which the branch was produced, the sap cannot rise in the leafless stump, and hence the bark gets dry and dies and the wound cannot be covered over.

C. Stems as Stores for Food in Reserve.

The leaves of a plant in active operation manufacture more food than can be used up immediately. The surplus of food is, therefore, laid up for any future emergency.

1. **Storage in Trees.**—The storage in trees is generally not preserved in special growths, but in numberless little cells along the way which the descending sap travels, *viz.,* in some cells of the inner bark, as also in certain cells extending from the cambium right to the centre of the tree, through the wood-tissue. These rows of cells, which are called *medullary rays,* appear as silvery rays when a woody stem is cut through transversely. The food in reserve is used by trees in different ways.

(a) Some trees, like the Teak, shed their leaves in the dry season. Most of the trees do the same during winter in cold countries. They begin to grow again with the return of the rains here, or of spring there, and are in a very short time *reclothed with* their beautiful *foliage,* as if by a miracle. This wonderful change is only made possible by the trees availing themselves of the ready-made food deposited in their stems, which is, when required again, carried to the buds where the plant has, as if by foresight, prepared the future leaves, and, in some cases, even flowers in miniature, before it dropped its leaves.

(b) At the time of *flowering* and *fruit-bearing* the consumption of food is so great that a previous storage of it is an absolute necessity. Some trees do not bring forth fruit every year, perhaps because the amount of food stored in one year does not suffice and they require several years' storage to produce their seeds. An instance of this kind is the shrub Strobilanthes growing on the slopes of the Ghauts and generally flowering after the lapse of seven or twelve years.

Other common examples of such *multiennials,* which like annuals die after fruiting, are the American Aloe (p. 156), the Fan Palm (p. 145), the Malabar Sago Palm (p. 145), and the Bamboo (p. 178).

2. **Storage in Herbs.**—We now know how trees lay up a store of food in their stems. There are also many herbs that are able to do it. Annual herbs may lay up some food until they produce seeds at the end of the season, when they consume the whole storage and die.

Biennials, such as Carrots, store up food in their roots, by which the flowers and fruits of the following year are nourished.

Perennial herbs, however, do not die at the end of a season or two, but let the parts which are above ground wither during the adverse season to sprout again from their underground parts when the rainy season reappears. These underground stems (rhizomes, bulbs, and tubers) have stores of food in them from which their buds derive their first nourishment when they burst into leaves.

4. THE FLOWER

We generally look at the flowers as bright and beautiful objects intended to be a source of pleasure to us; but they are created with a different purpose. Every living thing on this earth meets at one time or another with its destroyer, Death. To perpetuate its kind or species it is, however, endowed with the power of reproduction. This work is done in plants by their *flowers.* They produce seeds, from which, under favourable conditions, new plants of the same kind spring up. And we shall see that everything about the flower is subservient to this one aim.

Fig. 230. — Shoots arising from the edges of the leaf of Bryophyllum calycinum.

Many plants are able to propagate themselves also in other ways than by their flowers, as for instance the Hydrocotyle by runners, the Potato by tubers, the Onion by bulbs, the Lotus plant by the branches of its rhizome, the Bladderwort by detached segments of its stem, the Bryophyllum by sprouts rising from its leaves, etc. In all these cases some parts of the plant, other than its floral parts, detach themselves from the mother-plant and become new independent plants. This kind of reproduction is called *vegetative,* as compared with *sexual* reproduction by means of flowers and seeds.

A. The Parts of the Flower and their Structure.

1. **Parts.**—The flower is a short shoot with its leaves arranged in whorls or spirals. A complete flower has four whorls of leaves, which from outside to the centre are called sepals (forming the calyx), petals (forming the corolla), stamens, and pistils. (The stamens are often in two whorls, as in the Glory Lily, sometimes also in more than two whorls, as in the Lotus; sepals, petals, or pistils are sometimes also in more than one whorl.) If one of those four leaf-whorls is absent, the flower is called incomplete.

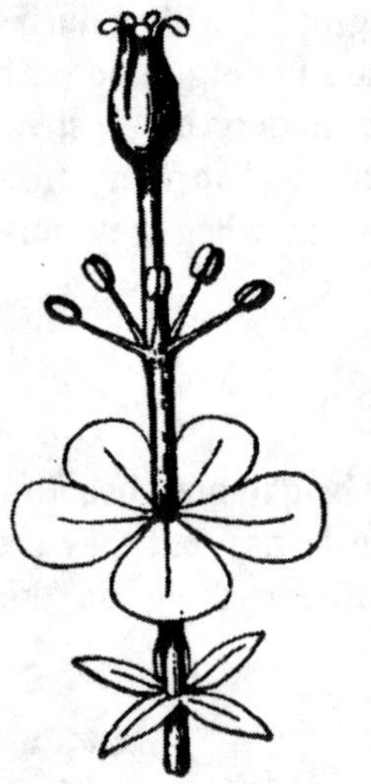

Fig. 231.—The parts of a complete flower: calyx, corolla, stamens, and pistil.

The function of each of the four kinds of floral leaves is different. The two external rings have to protect the internal ones in their unripe state. The corolla, serves as a means for the attraction of insects, inasmuch as it is vividly coloured. The two internal rings, again, are the organs of sexual propagation, the stamens producing pollen, and the pistils seeds. Hence the calyx and the corolla are the unessential, but stamens and pistils the essential parts of a flower.

2. **Incomplete Flowers.**—One or even both of the floral envelopes may be absent. Such a flower is called monochlamydeous, if there is only one envelope, as in Castor; and achlamydeous, if both are absent, as in Pepper. If one of the essential

parts is missing, it is unisexual; if both are present, bisexual. If on a plant bearing unisexual flowers both staminate and pistillate flowers are found, *e. g.*, Cocoanut, that plant is called monœcious (from Greek *monos* = one, and *oikos* = house), if, however, staminate flowers only are present on one plant of a species, and pistillate ones only on another, as in the Papaw, or the Palmyra Palm, the plant is called diœcious (*di* meaning two).

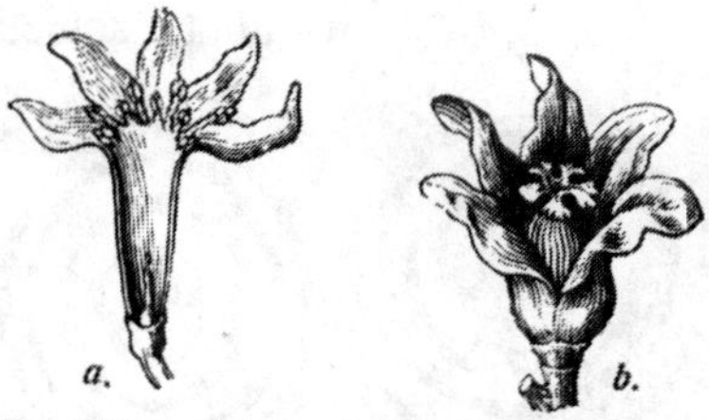

Fig. 232 — Unisexual flowers of the Papaw tree: *a.* staminate, *b.* pistillate.

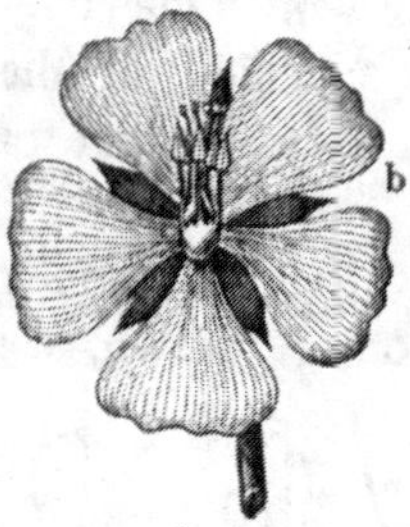

Fig. 233. — The radial flower of Flax.

3. **Symmetry of Flowers.** — If we compare several flowers as to the symmetry of their parts, we find two kinds. Some have their parts distributed so regularly round the axis that, divided by any diagonal section, the two halves are equal. Such flowers are radial. They are generally upright or hang down like bells. Examples are the Rose, the Shoeflower, and the Glory Lily. But if we can obtain two similar halves only by one diagonal section, the flower is laterally symmetrical or zygomorphic (from Greek *zygon* = yoke, and *morphe* = shape). Flowers of this kind usually open sideways, having their plane of symmetry vertical. This position is advantageous for their pollination, as the insects which are here the proper agents for pollination, come flying from the side. Examples of zygomorphic flowers are any of the Papilionaceæ, Labiatæ, Acanthaceæ or Orchidaceæ.

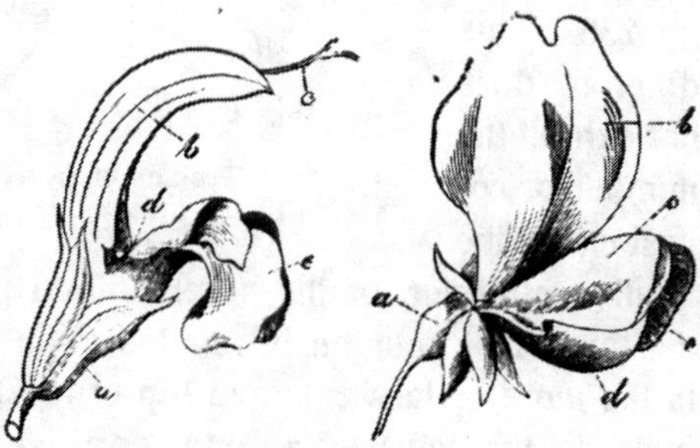
Fig. 234. — Zygomorphic flowers.

4. **Floral Diagram.** — Now, if we draw four or five concentric circles and mark the horizontal sections of the several floral leaves in their correct relative positions, we obtain something like the ground plan of the flower, and what in Botany is called the floral diagram of that plant. The conventional way in which the different parts should be drawn is shown in the adjoining illustration. As these diagrams show at a glance the number and relative position as well as the cohesion and adhesion of the different floral parts, they are very useful to a student of Botany.

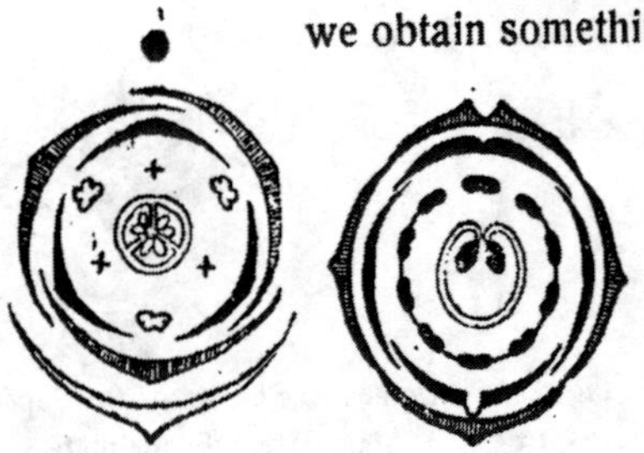

I. II.

Fig. 235. — I. Diagram of a radial flower (Iris). II. Diagram of a zygomorphic flower (Pea).

5. The **Receptacle.** — If a flower is cut longitudinally through its centre, and the parts of the section are drawn, we obtain something like the elevation of the structure of that flower, as shown in fig. 236. This diagram does not exhibit the number or cohesion, but the relative adhesion of the several floral parts, and, in particular, it shows their position in relation to the floral receptacle. This is the more or less enlarged top of the flower-stalk. Take, for instance, the flower of a Poppy or of a Custard Apple, and make such a longitudinal section. You will find that the sepals, the petals, the stamens, and the pistil, all rest on the slightly bulged surface of the receptacle. In the Custard Apple this receptacle

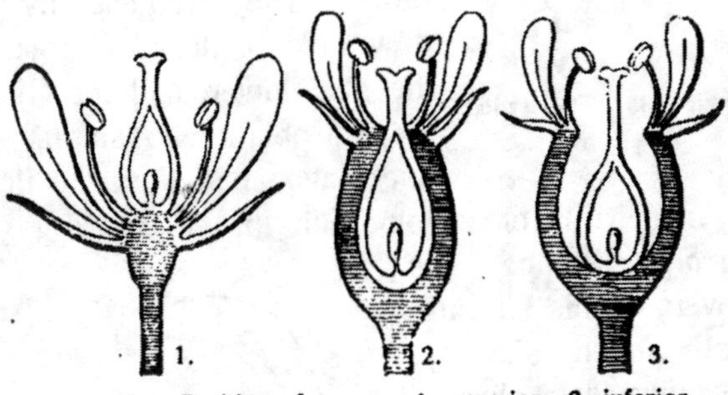

1. 2. 3.

Fig. 236. — Position of ovary: 1. superior, 2. inferior, 3. half-superior.

grows with the fruit and becomes the cone-shaped central part of the ripe fruit. The receptacle of the Sunflower becomes a large, flat disc supporting not only one, but a collection of flowers. Now, imagine the brim of this flat disc turned up so as to form a cup, and you have the shape of the receptacle, as it appears in a fig, supporting many small flowers, or in the Guava, the Rose, or the Poinciana supporting one flower only. These three latter flowers differ, however, as to the adhesion of the ovary to the receptacle, and as to the duration of the receptacle. In the Guava the receptacle adheres to the ovary so completely that even their suture cannot be discerned. It grows with the ovary, and becomes part of the pulpy pericarp of the guava fruit. The cup-like receptacle of the Rose contains several free carpels in its cavity, which do not adhere to the walls of the receptacle. The receptacle is also durable, and becomes the red, tough wall of the well-known Rose-fruit, attracting birds which scatter the seeds. In Poinciana, finally, the hollow receptacle is open and contains one free, stalked ovary. The receptacle takes no part in the formation of the fruit, and fades with the flower.

The Glory Lily and Crinum have small and flat receptacles. But the ovary is superior in Gloriosa, as the other floral leaves spring from the receptacle and below the origin of the ovary, whereas the ovary of Crinum is inferior, the ovary only originating from the receptacle and the remaining floral parts form the top of the ovary.

6. **The Floral Covers: Calyx and Corolla.**—If there are two distinct floral covers, the outer whorl of leaves, which are generally small and green, is called calyx, and the inner one corolla. The calyx is formed of sepals, the corolla of petals. In several flowers (Liliaceæ, Amaryllidaceæ, Orchidaceæ, Palmeæ) the two floral envelopes are coloured alike, and are about the same size: they are termed *perianth* as a whole.

The sepals of the *calyx* may be free (poly- or chori-sepalous), as in the Lotus, or connate = grown together (gamosepalous), as in Tulasi. As far as symmetry is concerned, the calyx may be radial or zygomorphic. The latter is often the case in Papilionaceæ and in Labiatæ. It is then termed bilabiate = two-lipped.

The calyx may also be spurred, and then it is often coloured, as in the Balsam.

We find the same conditions in the *corolla*. The flowers of the Lotus, the Mustard, the Shoeflower, the Rose, the Poinciana are choripetalous. Their petals have always more or less long stalks, called claws. Gamopetalous are the flowers of Ixora, Vinca, Tobacco, Argyreia, and Tulasi. Their crowns assume many shapes, as of a tube (the tubular disc-flowers of the Sunflower), or of a funnel (Bindweeds), or of a bell (Allamanda), or of a mouth with two lips (Labiatæ, Acanthaceæ). Choripetalous and zygomorphic are the butterfly-shaped flowers of the Papilionaceæ, gamopetalous and zygomorphic those of the Labiatæ. The flowers of the Convolvulaceæ are gamopetalous and radial, those of the Malvaceæ are choripetalous and radial. The two whorls of a perianth may combine and form one tube, as in Crinum and Eucharis Lily.

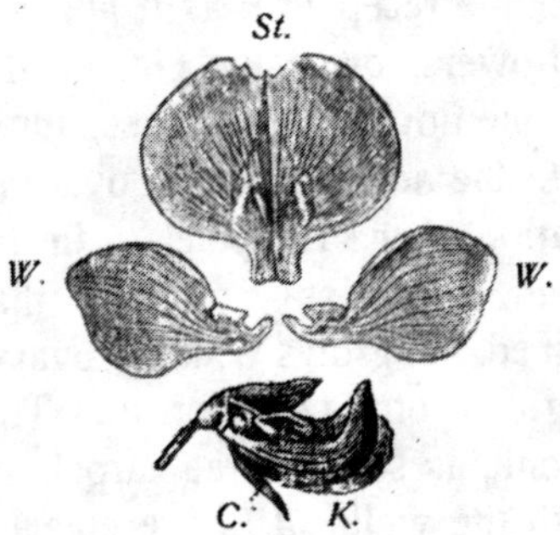

Fig. 237.—Flower of Pea dissected into its various parts: *St.* Standard. *W.* Wings. *K* Keel. *C.* Calyx of which the front part is removed.

In order to protect efficiently the inner essential organs, the sepals and petals do, as a rule, alternate, one with the other so that the gap between two sepals is closed by a petal, and inversely. In bud, moreover, these floral covers are variously arranged. The mode of their arrangement in relation to one another is called *æstivation*. This may be valvate (sepals of Shoeflower), or imbricate = overlapping (Poinciana, Rose), or contorted = twisted (Periwinkle), or plicate = folded (Argyreia).

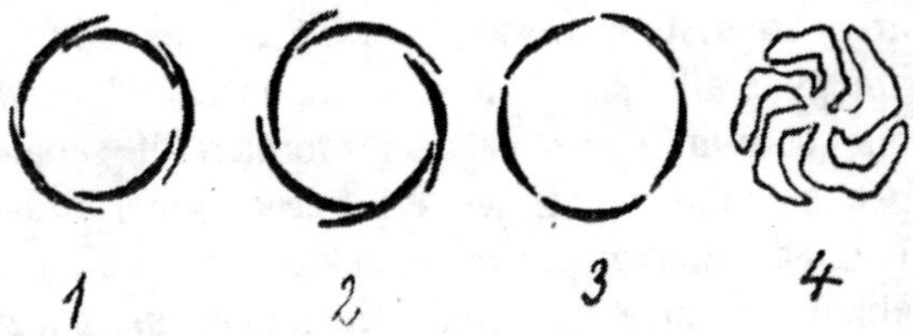

Fig. 238.—Æstivation of petals.

In some plants the calyx-leaves are dropped when the bud opens (Poppy), in other plants the corolla is shed similarly (Guava). The corolla may, however, be so constructed that it still affords the inner organs a good protection against the weather: In Adhatoda the stamens are covered under the upper lip of the corolla as under a shed; when the wind is strong, the blossoms of the Pea and other papilionaceous flowers turn their back to the wind, so that the inner parts are screened by the standard (see page 37); the long flower-tube of Sesamum is hanging and thus keeps the inner parts safe; the Water-lily closes up its petals during the rain, or at night, and opens them only to the sunshine.

The petals generally fade and fall after a short time, the sepals, however, are often persistent and remain on the fruit.

7. **The Stamens.**—*(a)* **Parts.**—The stamens consist of a stalk or *filament,* which supports two small bags, the *anther-bags,* containing a fine powder, the *pollen.* The anther is annexed to the filament by a small ligament, the *connective,* a tissue connecting the two anther-lobes.

(b) **Number, Adhesion** and **Cohesion.**—The stamens vary in number from one in Canna, two in Jasmine, three in Wheat, four in Ixora, five in Thorn-apple, six in Paddy, and so on, up to a hundred and more in Cereus.

As to their place of origin, they may rise directly from the receptacle (Anona, Guava), or from the tubular corolla (Ixora), or may be adnate to the pistil (Aristolochia).

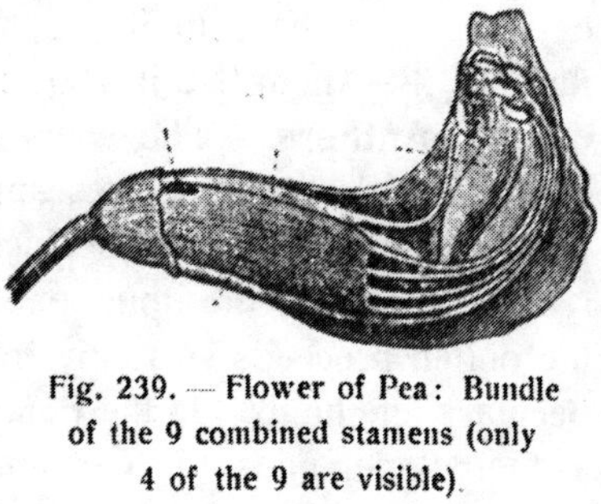

Fig. 239. — Flower of Pea: Bundle of the 9 combined stamens (only 4 of the 9 are visible).

With regard to their cohesion, they may be free from each other (Rose), or cohere into one bundle (Malvaceæ), or they may be in two bundles, as in the Papilionaceæ, where ordinarily nine filaments are combined (the anthers being free) and one stamen is entirely free; or they may be in several parcels as in Citron. In the compositæ the filaments of the stamens are free, but their syngenesious anthers cohere and form a tube.

(c) **Filament.**—The filaments may be long as in Crinum, or short as in Ixora, or absent as in Anona. Rarely the filament is flattened, as in Lotus. In the Labiatæ and in some Acanthaceæ there are four stamens in two pairs of different lengths; in the Crucifers we find six stamens, four longer and two shorter; in the Bilimbi tree five stamens are long and five short.

Fig. 240.—Flower of Compositæ: filaments (g) free anthers (e) united.

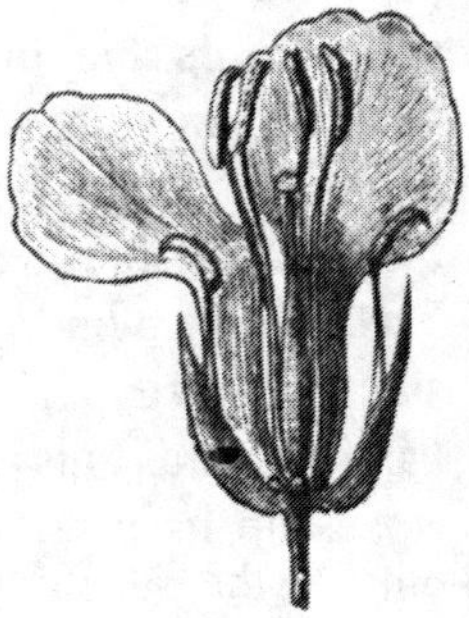
Fig. 241.—Flower of a Crucifer: four longer and two shorter stamens.

(d) **Connective.**—The connective is fixed to the filament usually by its base (basifixed), but in some flowers, as in Gloriosa, by the middle of its back (dorsifixed). In the latter case the filament becomes very thin at its end, and the heavy anther can freely swing on it. This is advantageous to the plant which is visited by night-moths (see pollination). Usually, the pollen-bags are attached to the broad frontal sides of the connective; in the Labiatæ, however, they are placed at their ends (see Salvia, page 106). In some flowers, again, the connective is produced beyond the anther; in the Compositæ it ends in a triangular flap; in the Anonaceæ in a round cap.

(e) **Anthers.**—The ways in which the anther-bags open (dehisce) to let the pollen escape, are also different in different plants. Commonly they dehisce in two longitudinal rents, the rent beginning at the upper end. The slits become so wide and the pollen proceeds so richly from the two bags that the pollen becomes one lump. In the Potato the anther-bags open by pores at their upper ends; in Cinnamon by valves. In the Orchises the pollen coheres and forms a dense mass.

Finally, it may be remarked that the anthers in some flowers open towards the centre (Sunflower), and in others towards the outside (Anona). In the former case they are called introrse, in the latter extrorse.

(f) **Pollen.**—The colour of the pollen changes from bright yellow to orange or red and, in some plants, it is even black.

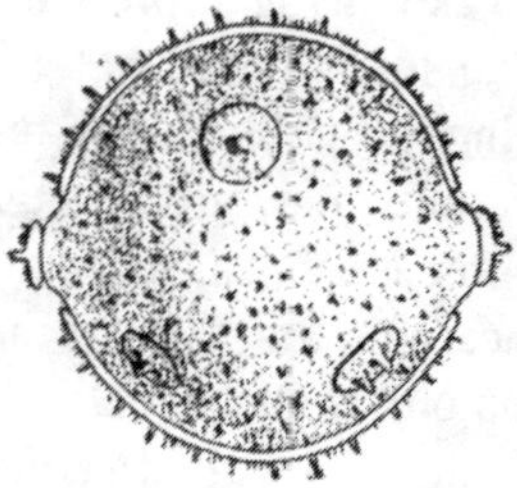

Fig. 242.—A grain of pollen of Cucurbita maxima (magnified 480 times).

Fig. 243. Fertilization of ovule.

If we put pollen-grains into water, they absorb so much of it that they burst. The same happens, if they are exposed to rain or dew while on the flower. Hence, the pollen is protected from damp in many flowers. If we sprinkle some pollen-grains over a drop of very weak sugar and water, they do not burst, but throw out long threads. This is what they also do when they are brought on the sticky end of the pistil. The thread grows downward between the tissue of the pistil until it comes to an ovule, which is then fertilized.

8. **The Pistil.**—

(a) **Parts.**—The chief part of the *pistil* is the seed-box or *ovary,* containing tiny seeds or rather ovules, which are destined to become seeds, when the fruit ripens after fertilization. This can only happen when the pollen is brought into contact with the ovules. Hence, the ovary tapers at its upper end into a more or less slender pillar, the *style.* To enable the style to hold fast the pollen-grains that happen to fall on it, its end, the *stigma,* is provided with tiny warts or hairs, which sometimes make it look like velvet (Shoeflower), and also with a sticky liquid, which exudes from the surface of the stigma. Some stigmas (Paddy) are feathery. Some flowers, like the Lotus, have no style; the stigma is then said to be sessile.

(b) **Ovary.**—The ovary is made up of one or several fruit-leaves or carpels (from Greek *karpos* = fruit). The line in which

the edges of a carpellary leaf are joined is termed its ventral suture. It is there that usually the ovules are attached alternately, one right, the other left. The mid-rib of the carpellary leaf is its dorsal suture. Such is the structure of an ovary that consists of a single carpel, as the Pea. Take any leaf, preferably a lanceolate one, bend its edges over the mid-rib so that they come together, and you produce something not unlike a pea pod. If there are several free or *apocarpous* carpels, they are arranged either radially or spirally on the receptacle (Vinca, Clematis, Uvaria). In many cases, however, the carpels are connate and form a *syncarpous* ovary, which contains one cavity or cell (also called *loculus*). Here the two edges of each carpellary leaf do not join together, but touch those of the neighbouring carpels, right and left, as in the Orchis or also in the Poppy. If the edges of each carpel do join one another, they meet all in the centre of the ovary, which thus becomes divided into two (Chilli), or three (Gloriosa), or as many cells or loculi as the ovary is composed of carpels. The ovules are usually attached to the edges of the carpellary leaves, where a distinct growth, called the *placenta,* is formed. In a many-celled ovary the placentation is axile and in a one-celled ovary that is composed of several carpels, it is usually parietal (from Latin *paries* = wall). In a few cases, however, the seeds are not attached to the edges of the carpels, but to a pillar-like growth in the centre of the ovary, thus forming a central placenta (*e. g.,* in Antigonon, Basella).

1. 2. 3.

Fig. 244. — Structure of ovary: 1. The ovary consists of one carpellary leaf (Pea). 2. It is composed of 3 carpels (Gloriosa). 3. Many carpels form the ovary (Poppy).

B. The Function of the Flower: Pollination.

1. Various Provisions for Cross-pollination.

Most flowers have stamens and pistils growing together on the same flower. It does not follow from this, however, that the pistil of such a flower is pollinated from its own stamens. This would be called self-pollination. Although cases of this kind do occur in nature, they are not at all common. It has been proved by many careful observations and experiments that plants on which the flowers have been self-pollinated, bear poor and insignificant fruit. As a rule, pistils are pollinated by pollen from other flowers, and to ensure such **cross-pollination** many wonderful and interesting arrangements exist, some of which we shall now consider.

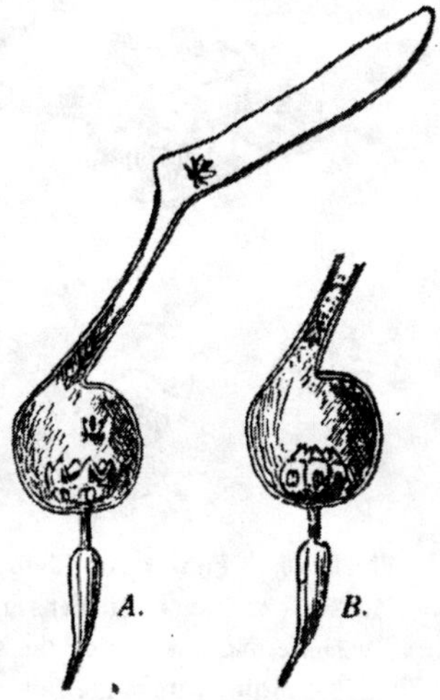

Fig. 245. — Aristolochia. *A.* Stigma mature before ripening of stamens. *B.* Pollen developed after fading of stigma.

(a) Stamens and pistils are *distributed over different flowers* (monœcious and diœcious plants).

(b) In some plants, where the stamens and pistils do occur in the same flower, these organs *mature at different times:* in the Sunflower the stamens open when the stigma is still undeveloped; but in the Aroideæ and in Aristolochia the stigma is mature before the stamens develop their pollen.

(c) If both mature at the same time, the stamens and pistils are sometimes *so placed that the pollen cannot easily reach the stigma* of its own flower (Hibiscus).

(d) The flowers of the shrub *Clerodendron infortunatum (Kan.* Iṭṭēvu; *Mal.* Peragu) exhibit a remarkable contrivance for cross-pollination. The white corolla and the strong, sweet scent make the flower conspicuous at dusk to moths which stretch

their long tongues into the flower-tube to obtain the nectar hidden in its depth. While thus hovering in front of the flower they touch the anthers of the four stamens that protrude from the flowers. The *style* of these flowers is *bent back* in a long downward arch. The moth will then move on to another flower which may have opened the previous night and now has its stamens curled up but the *style straight,* so that the moth cannot help touching and thus pollinating it with the pollen of the flower from which it came.

1.

2.

3.

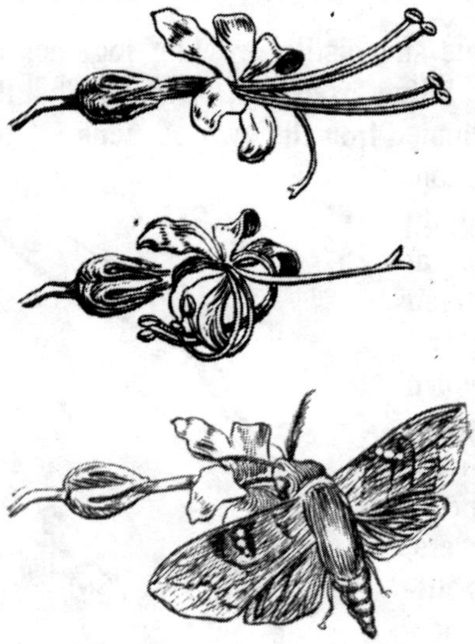

Fig. 246.—Flowers of Clerodendron.
1. First position of stamens and style.
2. Second position of the same.
3. Moth visiting the flower of Clerodendron.

(e) We have found another contrivance to avoid self-pollination in the Orchids (page 161), where the pollen-masses are in separate pouches, and can only be removed by a bee which carries them to another flower to pollinate it.

As cross-pollination is advantageous to the plant, and as the plant is unable to move, it requires some assistance to carry the pollen from one flower to another. This is obtained from insects or from the wind.

2. Pollination by the Agency of Insects.

(a) **What the plants can offer to their guests.**—The conveyance of the pollen from one flower to another is not done by animals purposely or voluntarily. They pay the plants visits for their own sake only. But while obtaining some benefit from the plants they, in their turn, unconsciously render them a very

useful service in carrying their pollen to the next flower. What they seek and find in the flowers is first of all a sweet juice, called nectar. (It is not honey, but becomes honey in the body of the bee.) This sweet liquid is secreted by certain glands forming the *nectary* which may belong to any of the floral parts, calyx, corolla, stamen, pistil or receptacle (*cf.* Gloriosa, Mango, Aristolochia, Bindweed). In addition, there are often streaks or dots upon the corolla pointing the way to the nectar; such *nectar-guides* are seen, for instance, in the Mango. The nectar is generally hidden away in the depths of the flower, and the insect must, therefore, either be furnished with a long tongue, like butterflies and moths, or must actually get its body right into the corolla, like the bee. The humble bee, for instance, when visiting a labiate flower, alights on the lower lip of the corolla, which is admirably suited for a landing place. The weight of the insect naturally bears the flower down, and, as the bee pushes its body into the throat of the flower to reach the nectar at the bottom, its back comes in contact with the anthers and rubs off some of the pollen. This is unconsciously carried to the next flower visited, and some is rubbed off by the style.

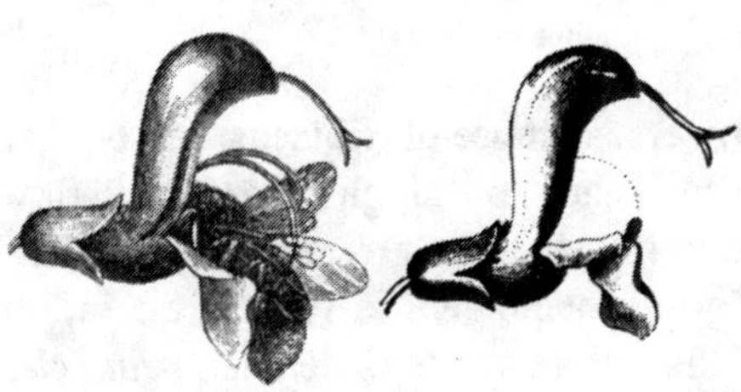

Fig. 247.—Bee visiting the flower of Salvia.

Besides nectar, many flowers offer them their *pollen* as food. Several flowers possess no nectar at all, but instead of it plenty of pollen in their numerous stamens (Poppy, Rose). These flowers are erect and have the shape of shells or cups, so that the falling dust may not be spilt and lost (page 8).

Some flowers offer their visitors nice and snug homes to live in for a while (Aroideæ, Aristolochia); and the Banyan tree even allows the small midges that can enter the little openings of the figs, to lay their eggs in the figs (page 118).

(b) **How the plants attract visitors.** (i) Colours.—The petals of the flowers are usually brightly coloured, and are

readily seen from a distance. If the corolla is inconspicuous, the bracts may become coloured, as in Bougainvillea. In Mussænda (*Kan.* Bellotti; *Mal.* Vellila) one calyx-lobe is much enlarged and looks like an ordinary leaf, but is white. Flowers that open in the evening and must, therefore, be pollinated by night-moths, are generally white or pale, so as to be easily seen in the twilight (Jasmine, Crinum, Clerodendron, Nyctanthes, etc.).

Fig. 248. — Mussænda frondosa. The large white leaf is one of the five lobes of the calyx.

(ii) Inflorescences. — Large flowers, as those of Hibiscus, can be discovered by visiting insects even if they are single; but small flowers become conspicuous only if they are grouped together in greater numbers. Such a group of flowers is termed an inflorescence. The main axis of the inflorescence is denoted *peduncle,* and the stalks supporting the flowers *pedicels.* The leaves, from the axils of which the pedicels issue, are smaller than the foliage leaves of that plant, and are called *bracts.* The epicalyx of the Malvaceæ (p. 13), the involucre of the Compositæ (p. 65), the spathe of the Palmeæ and Aroïdeæ (p. 140 and 147) and the glumes of the Gramineæ (p. 173) are modifications or specialized forms of bracts.

There are a great many of various inflorescences, which, however, can be referred mainly to two types. In one type, the racemose or indefinite type, the main axis is stronger than any of the side axes; in the other, the cymose or definite type, the main axis is shorter and weaker than the side axis. Thus the racemose inflorescence is a monopodium, and the cymose inflorescence a sympodium (see page 245).

(aa) *Racemose inflorescences.* — Here the principal axis goes on elongating and gives off branches bearing flowers. The flowers at the lower part of the peduncle are first developed and, therefore, open earlier than the upper ones. (This is shown in the illustrations by the various sizes of the circles denoting the flowers.)

If the peduncle bears stalked flowers, the inflorescence is called a *raceme* (Poinciana), if the pedicels are branched again, the raceme becomes a *panicle* (Mango, Paddy). If the flowers of a raceme are sessile on the peduncle, we have a *spike* (Pepper Vine, Orchis). A spike with a fleshy axis, covered by a spathe, is a *spadix* (Aroideæ) but one with a weak pendulous axis a *catkin* (Sapium).

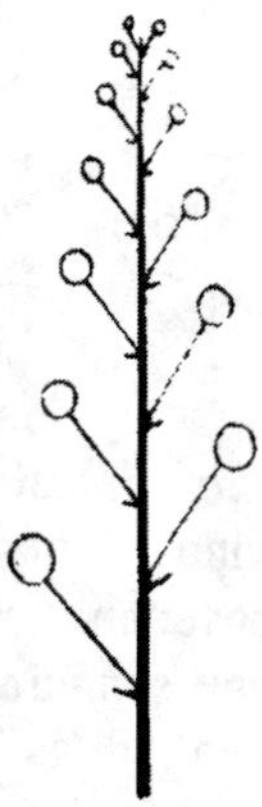

Fig. 249. Raceme.

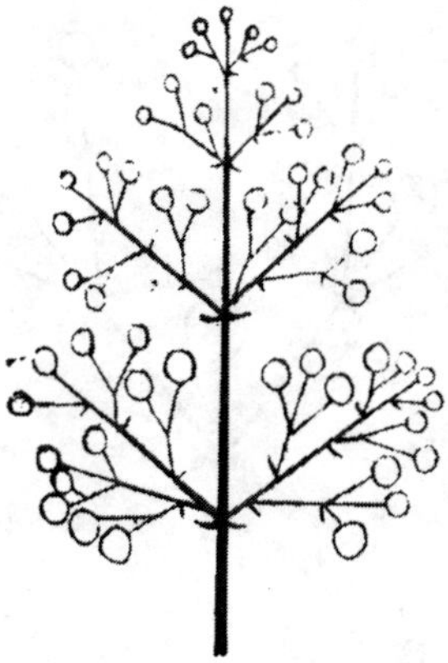

Fig. 250. Panicle.

Fig. 251. Spike.

Racemose inflorescences are also the *umbel* and the *head*. Here the main axis is shortened. The umbel has its flowers stalked and all springing from the top of the peduncle. In the head of the Compositæ the peduncle ends in a more or less flattened receptacle surrounded by an involucre of bracts, and supporting densely crowded, sessile flowers.

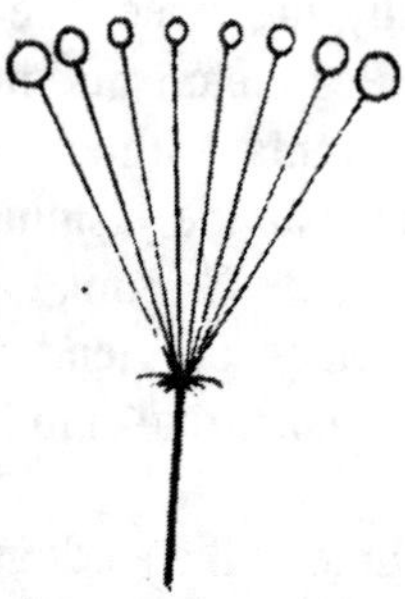

Fig. 252. — Umbel.

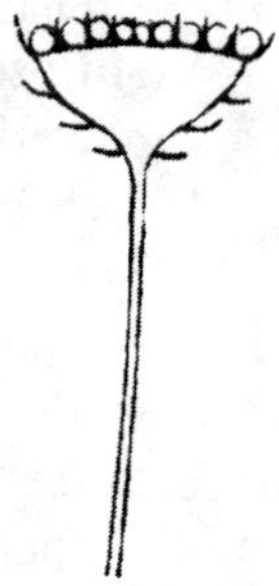

Fig. 253. — Head.

(bb) Cymose inflorescences. — The main axis terminates in a flower and does not elongate. But below this flower, which opens first, one or several lateral pedicels are given off, each terminating again in a flower. This may be repeated several times. Thus an inflorescence is formed, the axis of which is composed of the consecutive axes of the first, second, third order and so on. Figures 254—256 represent cymes in which the successive lateral axes come off either in one and

the same direction or in alternate directions. Cymose inflorescences may be studied in Ipomæa, Clerodendron, Ixora, and in the Labiatæ.

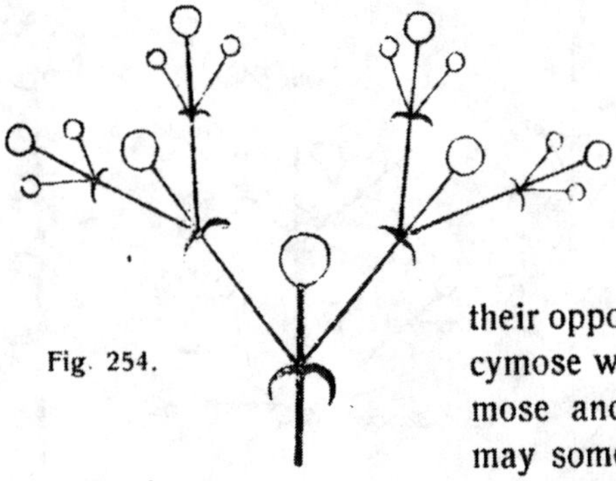

Fig. 254.

In Ixora the pedicels are much reduced, so that a *fascicle* of flowers is formed. The Labiatæ have small fascicles in the axils of their opposite leaves, thus forming cymose whorls or *verticels.* Racemose and cymose inflorescences may sometimes be found in combination, as for instance, in Clerodendron infortunatum.

(iii) SCENTS.—Strong and various scents are also great helps to attract insects. The bee-tribe and butterflies are specially attracted by the sweet scent of Roses, Peas, etc.; and the powerful scents emitted by such flowers as the Jasmine, Tobacco, and Crinum, as evening comes on, tend to guide the nocturnal moths to them.

Fig. 255.

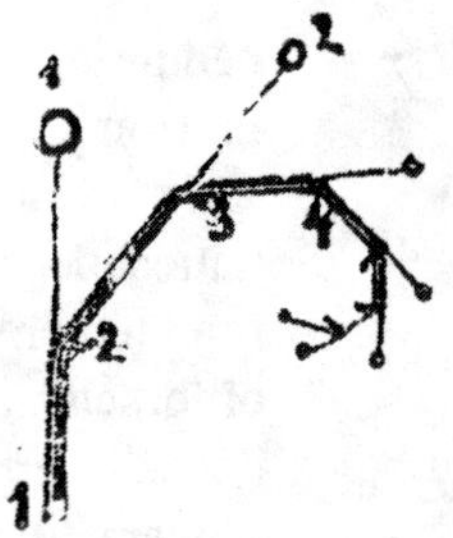

Fig. 256.

Figs. 254—256 represent cymes

Sometimes the odours used to attract insects (flies) are the reverse of pleasant to us. As an instance the Nux-vomica tree may be mentioned, or *Aristolochia gigantea* (*Kan.* Bātkōḷi hūvu).

Scentless flowers usually have some equivalent form of attraction, such as gaudy colours, abundance of pollen, or the grouping of a number of small florets; whereas inconspicuous flowers (Violet) are often endowed with a particularly strong smell.

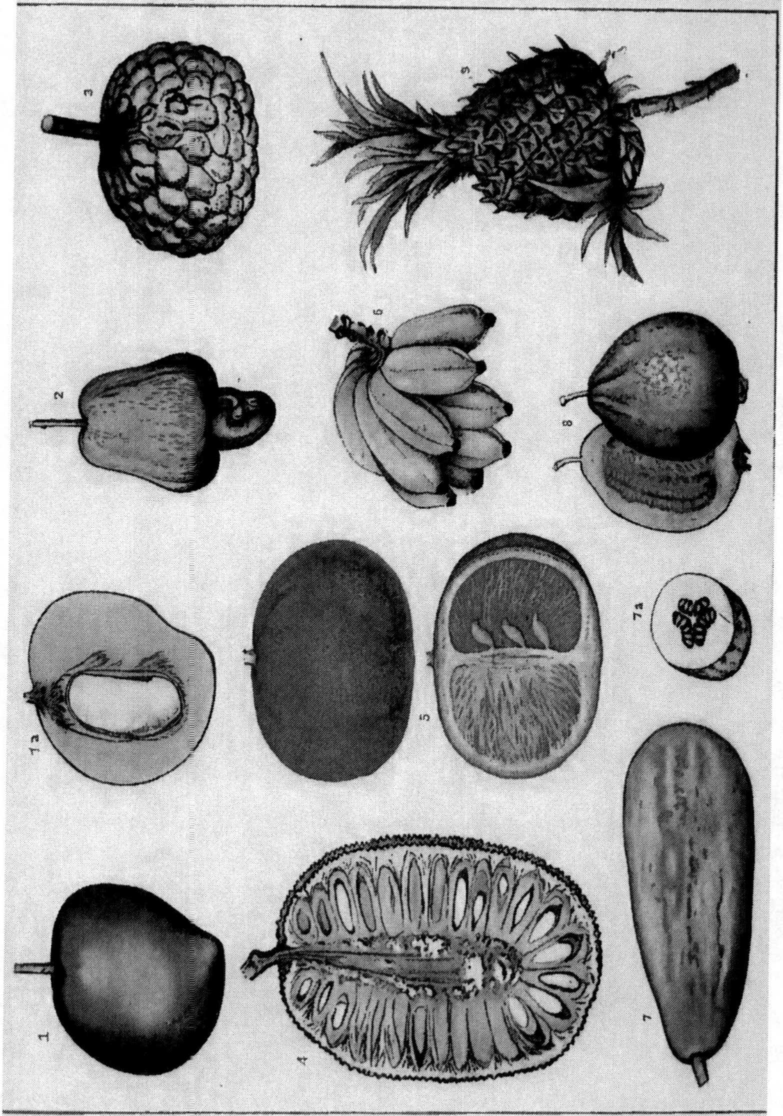

Fig. 257.—Fruits.

3. Pollination by the Agency of the Wind.

Among the plants whose flowers are pollinated by the help of the wind, the chief are the Grasses. The principal characteristics by which such plants may be recognised are:—

1. The flowers are *inconspicuous, scentless* and have *no sugary secretions.*

2. Their *anthers hang out of the flower,* so that the pollen can be easily set free and carried away by the wind.

3. The *pollen* is produced very *abundantly* to allow for wastage.

4. The *pollen-grains are dry* and *small* and, therefore, *light.*

5. The *stigmas are large* and *feathered,* so that pollen-grains floating in the air are easily caught.

5. THE FRUIT AND THE SEED

A. How fruits are formed.

When the ovules are fertilized, the ovary begins to grow and becomes the fruit, consisting of the pericarp and the seeds. If the ovary is adnate to the receptacle, as in the Guava and the Rose, the latter is also affected by the change and becomes a part of the fruit cover. The same happens in the so-called *collective* or *multiple fruits* (Mulberry, Jack, Fig, Ananas) that consist of inflorescences of densely crowded flowers (or fruits) on a common receptacle. Style and stigma usually take no part in the changes caused by the fertilization of the ovules. They fade and are dropped with the rest of the fading flower. In the Poppy, as an exception, the sessile stigma remains on the top of the ripe fruit. The calyx is often persistent. So in the Leguminosæ, in the Myrtaceæ and in the Labiatæ. In many of the Compositæ the calyx forms the pappus, a flying apparatus, and in Physalis an inflated cover over the juicy berry.

1. **The Pericarp.**—The covering of the fruit (pericarp) is, with the modifications mentioned above, formed of the carpellary leaves. Three parts are distinguished in it.

The outer part *(exocarp)* may be smooth and shining, as in the Chilli; or glaucous, as in the Mango; or covered with stinging hairs, as in Mucuna pruriens (*Kan.* Nāyisoṇagu; *Mal.* Nāyikuruṇa); or prickly, as in the Thorn-apple. All these qualities serve as protections from damp or from the aggressions by animals, or they are a means for the attraction of animals which would scatter the seeds.

The middle part *(mesocarp)* is fibrous in the Cocoanut, but soft and juicy in many other fruits. The pulpy mass remains, however, sour or bitter during the period of growth and sweetens only when ripe. This quality, likewise, serves as a means for the scattering of the seeds by animals.

The inner part *(endocarp)* in these fruits is hard like stone, called pyrene, and forms a very strong protective covering to their seeds. In other fruits it forms a skinny membrane lining the inner cavity of the fruits in which the seeds rest, or the partitions of the fruit, as in the Orange.

The Seed.—While the seeds grow and require nourishment, they are attached to the placenta by a cord, called *funicle*. When they are mature, the funicle withers, and a scar, called *hilum,* is left on the seed.

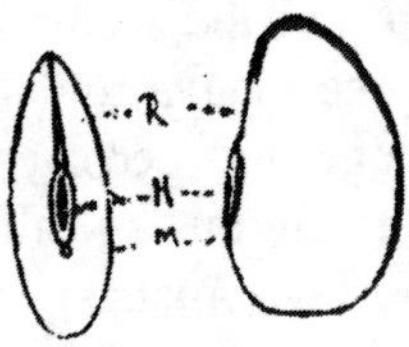

Fig. 258.—Seed of Horse gram (enlarged).

The seed consists of the embryo and its protective coats. The outer coat, called *testa,* has, by the side of the hilum, a small hole, the *micropyle,* by which the pollen-tube entered the ovule for fertilization (page 259). The testa is usually hard, sometimes possessing projections into the interior ("ruminating testa") as in the Nutmeg and in the Custard Apple, sometimes developing appendages for dissemination, as in Madar, Bignonia, and Lagerstrœmia. In the Pomegranate it is juicy. A growth of spongy tissue is found on the testa of certain seeds. If it appears round the micropyle, as in the Castor-oil seed, it is called *caruncle;* if it originates from the attachment of the funicle, it is termed *aril* (Nutmeg).

The inner coat, called *tegmen,* is absent in many seeds. It may be seen in the Castor-oil seed.

The *embryo* is the young plant, consisting of a short stem *(hypocotyl)* with seed-leaves *(cotyledons)* and a minute bud *(plumule)* at its upper end, and a minute root *(radicle)* at its lower end. The hypocotyl remains short in the Mango and in the Pea, but lengthens out in the French Bean, Horse gram, Castor, Cucumber, and many other seeds, when they germinate. The cotyledons may be thick and stored with plant-food (Horse gram, Bean), or thin (Castor). They may also be flat (Bean, Castor), or folded (Ipomæa).

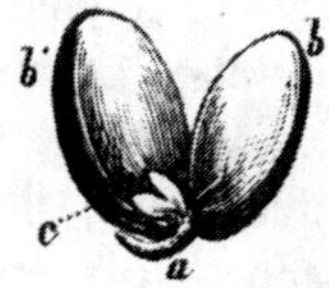

Fig. 259. — Seed of the Bean: *a*. Radicle. *b*. Cotyledons. *c*. Plumule.

The number of the cotyledons marks two large classes of the flowering plants, *viz.*, the dicotyledons and the monocotyledons. In the monocotyledons the cotyledon has a cylindrical shape sheathing the stem and its bud, and developing at the side adjoining the endosperm, of which we shall speak presently, into a shield-like structure as in Maize, or into a globular body as in the Cocoanut. Here, as well as in many dicotyledonous plants, the cotyledons have the function of absorbing a food-substance, which forms a separate tissue within the seed coats, and that is the *endosperm*. (See Germination of Castor seed, of Rice, and of Cocoanut.) Seeds containing endosperm are termed

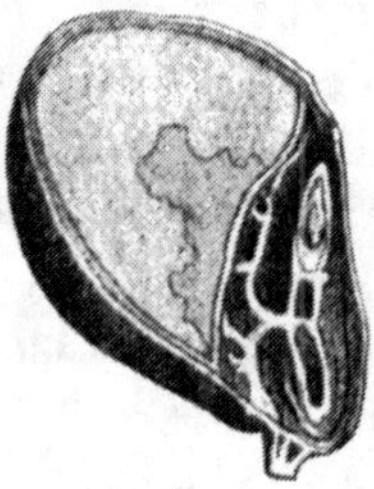

Fig. 260. — Endosperm in the seed of Maize.

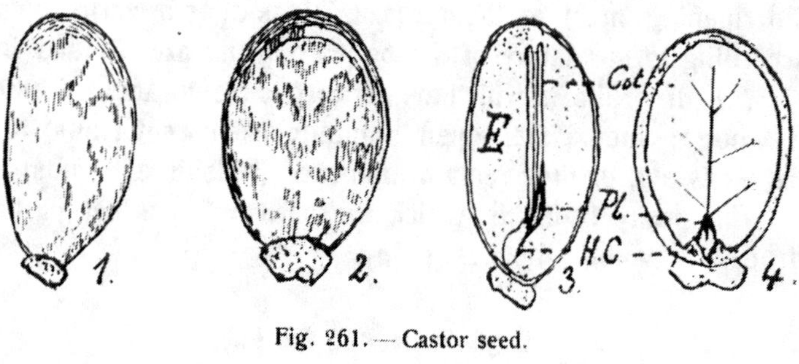

Fig. 261. — Castor seed.

endospermous, and those in which endosperm is not present, exendospermous. Exendospermous seeds have the food-substance necessary for the growing seedling packed in their cotyledons, which thus become swollen. The food-substance either contained by the cotyledons or by the endosperm is also utilized for food by man, as well as by many animals.

When the seeds are mature and their germs in a position to live alone, they must leave the protecting covering (pericarp) and be dispersed, so as not to germinate just under or near the mother-plant where the young plants would simply suffer from her shade, and from the fact that she had used up some of the supplies of mineral food available, and that they would have a hard struggle for life with one another. The germs or young plants must, therefore, be so constructed as to be able to go on a long journey without perishing. The are in this state in the ripe seed. When fully formed, they cease to be watery, the place formerly occupied by water being now filled with starch or oil, and the seed-coat becomes hard.

Many seeds, it is true, do not end their journey in a very suitable place, but die by hundreds and thousands for want of congenial surroundings, and it is for this reason that such large numbers of seeds are produced.

B. How fruits allow their seeds to escape.

In order to enable the seeds to start on their journey, the fruit which had to protect them during the period of their growth and ripening, must set them free. This is done in various ways. According to the nature of the pericarp fruits are divided into:

(a) dry fruits, in which the pericarp is membranous, leathery, or woody; they either open naturally (dehiscent fruits) to let the seeds out, or they remain unopened (indehiscent fruits)

(b) fleshy fruits, in which the pericarp is either partially (drupes) or wholly (berries) pulpy.

Dry dehiscent fruits.

The simplest form of this kind is a capsule originating from a single carpel and bearing the seeds at its ventral suture. If such a pod dehisces in one slit and in its ventral suture, it is called a **follicle.** Sometimes such follicles are doubled, as in Vinca originating from two apocarpous fruit-leaves. If a one-carpelled pod slits both in its ventral and dorsal sutures, as in the Horse gram, Pea, and in most Leguminosæ, it is called a **legume.**

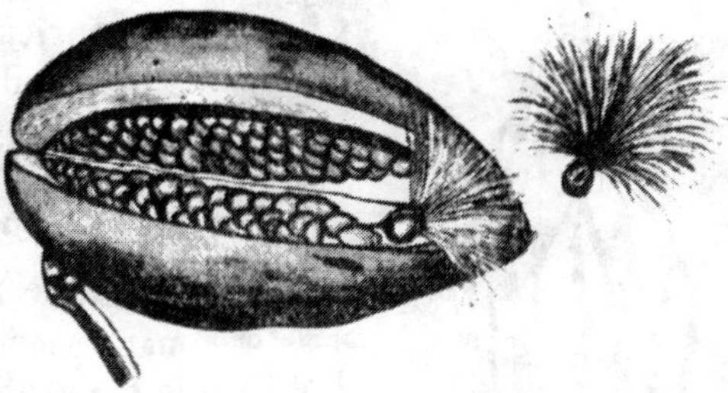

Fig. 262. — Follicle of Madar.

Crucifers have **siliques** consisting of two carpels, divided by a false septum into two loculi, the seeds being borne by the frame of that septum. All other

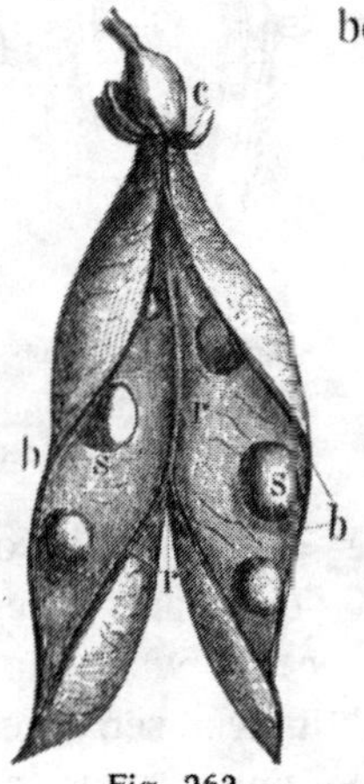

Fig. 263. Legume of Pea.

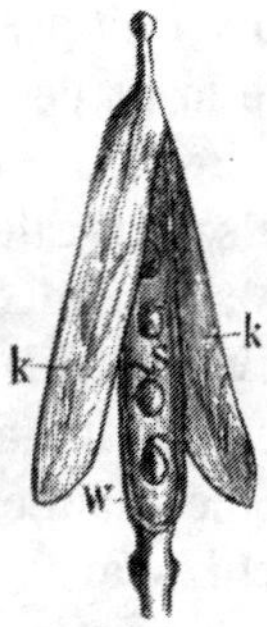

Fig. 264. Silique of a Crucifer.

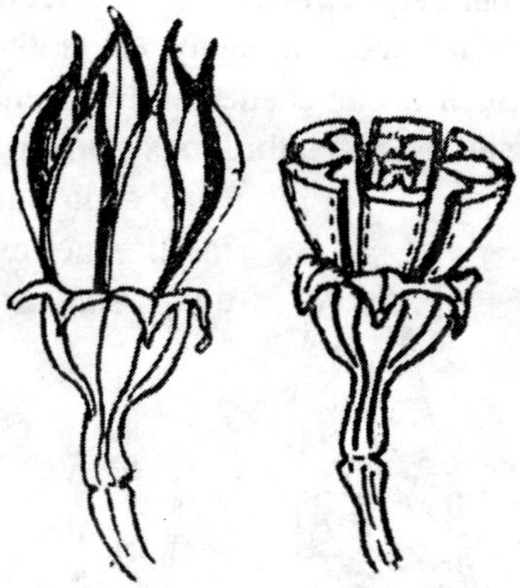

Fig. 265. — Loculicidal capsule of Lagerstrœmia.

dry dehiscent pods that cannot be classed under follicles, legumes or siliques, are simply called **capsules.** They are distinguished according to the number of cells or loculi they contain and to their mode of dehiscence. Commonly, they dehisce in longitudinal slits beginning from the top and dividing the capsule into so many open parts as the capsule contains cells. If the line of dehiscence is along the mid-rib of the carpels, the dehiscence is **loculicidal,** as in Lagerstrœmia; if it is along the partition walls (septa), as in Aristolochia and Gloriosa, **septicidal.** In the Poppy and Osbeckia the capsules open by pores at the top: they are **poricidal;** and in Portulaca a lid is detached from the pod by **circumsciss** dehiscion.

Fig. 266.—Septicidal capsule of Aristolochia.

Fig. 267. Poricidal capsule of Poppy.

Indehiscent dry fruits.

These may be one-seeded or many-seeded. A one-seeded fruit is the **grain** of Rice or Maize: in this the thin pericarp closely adheres to the seed. In the **achenium** or **nut** the seed is free from the pericarp, which may be leathery as in the Sunflower, or stony as in the Hazelnut. If such a one-seeded indehiscent dry fruit possesses a membranous appendage to its pericarp, as in Butea frondosa or in Ailanthes excelsa, it is called a **samara**. Many-seeded dry indehiscent fruits are characteristic of some families. They are called **schizocarps,** because when ripe each carpel with its seed separates from its neighbouring carpels, as in Sida (Malvaceæ) and in Tulasi (Labiatæ).

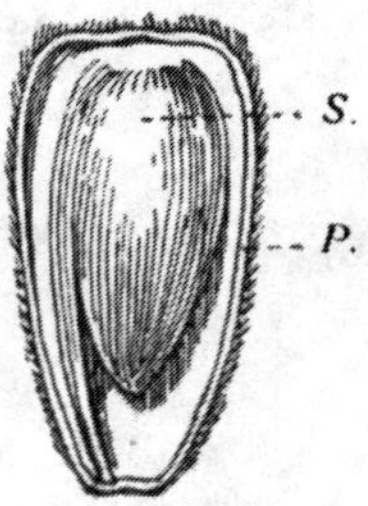

Fig. 268. Achenium of Sunflower (opened). *P.* Pericarp. *S.* Seed.

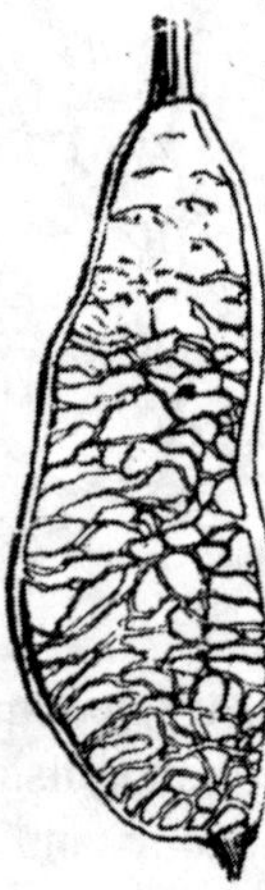

Fig. 269.—Samara of Butea frondosa.

Fig. 270. Schizocarp of Tumbe. (Front part of calyx removed.)

Fleshy Fruits.

Such are the **drupes,** in which the seed or the seeds are enclosed in a hard shell each. The Mango or the Cocoanut is a one-seeded drupe, the Coffea fruit a two-seeded one, and the fruit of the Palmyra palm a three-seeded one.

Berries have no stony endocarp, and their seeds are enclosed directly by the pulp. The number of the carpels forming the berry is two in Chilli, three in Banana, five in Papaw. The Chilli and Banana have axile, and the Papaw parietal placentation.

C. How the seeds are dispersed.

It is not enough that fruits should allow their seeds to escape. If they fall down and begin life again *directly under* the leaves of the parent plant, they cannot, as already stated, get sufficient light and air, but finding only a soil from which a great deal of the plant-food had already been extracted by the mother-plant, they must starve. To ensure the dispersion of the seeds over a wide area, various wonderful provisions are made. As the plants cannot move of themselves, they often make use of the agency of running water, of wind, animals and birds, and with their help undertake long journeys to distant countries, and even cross oceans.

Fig. 271.—Capsule of a Balsam: 1. closed, 2. exploding and scattering the seeds.

(a) **Dispersed by mechanical contrivances.**—Some plants contain within themselves the means necessary to scatter the seeds. This is in some cases a spring apparatus by which the seeds are thrown away from the parent plant, as in the capsules of the Acanthaceæ or in the Balsams. "The seed-pod is generally in a state of tension, due to the gradual drying up of the tissue. Then a puff of wind, a slight blow, or even a change in the atmospheric condition of the air, gives the final impetus, causing the pod to burst with such force that seeds are thrown out in all directions" *(Brightwen)*.

In other cases, the seeds are furnished with *awns,* as in many Grasses. Each awn is thickly set with bristles which allow the awn to move only in one direction. As these are sensitive to moisture, the difference in the amount of moisture in the air lengthens and contracts them, so that the seeds attached to them are slowly but surely drawn away many inches.

(b) **Dispersed by water.**—Many water-plants have seeds or fruits which float. An air-bubble is attached to the seed of the Water-lily which causes it to rise to the surface. The cocoanut is provided with a strong, but light covering, and, if it falls into the sea, it may be carried by the waves hundreds of miles. How many seeds that fall to the ground during the hot season are washed away by the torrents of rain of the bursting monsoon and landed on a spot far away from the mother-plant, the hard seed-shells protecting the tender germ within from destruction!

(c) **Carried by the wind.**—The winged fruits of so many trees are very well known, such as Hopea Wightiana (*Kan.* Karmara; *Mal.* Ilapoṅgu; *Tam.* Koṅgu) or Butea frondosa, in which the calyx, or the pericarp is enlarged and

Fig. 272.—Winged fruit of Hopea Wightiana.

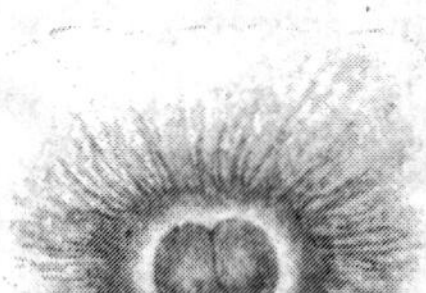

Fig. 273.—Winged seed of Stereospermum.

becomes a wing in the fruit. In Stereospermum (*Kan.* Puruḷi; *Mal.* Pātiri; *Tam.* Pādiri) the winged appendage belongs to the seed. The fruits of many Compositæ and the seeds of the Asclepiadaceæ are provided with silky or feathery hairs (called "pappus" in Compositæ), which catch the wind and allow the seed to be borne away even by the gentlest breeze. In some cases, as in the Poppy and Ladies' Fingers, the capsules are placed on long stalks, which the wind shakes with so much force that the seeds are thrown away as from a sling. In the Orchids the seeds, which are minute and light, are scattered also by the wind.

Fig. 274. — Seeds of Vernonia dispersed by wind.

(d) **Dispersed by animals and birds.** — This is done either unconsciously or consciously: *unconsciously*, when the seeds by their hooks and bristles (*e. g.,* in Triumfetta, or in Urena lobata) become attached to the skin of animals, or by the sticky mass of their pulp (Loranthus) to the beaks of birds which carry them away: *consciously,* when animals or birds, attracted by the bright colour and the fragrant smell of succulent berries and drupes, revel upon the sweet, soft flesh, but reject the seeds (Mango). If they swallow the seed also, the latter is enabled to resist the action of digestion by its hard covering and passes through the animal undigested (Coffee, Fig).

D. How the seeds produce new plants.

When a seed has thus finished its journey and has found favourable conditions (warmth, moisture, and air) on the spot where it has finally settled, it awakes from the state of rest in which the germ contained in it has tided over a season that might have been fatal to its life, and begins to germinate. First it absorbs moisture through its skin. This, combined with warmth and the oxygen of the air, sets up a change in the condition of the seed. The germ swells, breaks the seed-skin, and begins to grow: the tiny root of the embryo lengthens and grows downwards, while the hypocotyl stretches upwards. For this first process of growth the young plant requires the food-store provided by the mother-plant. It can be observed in every germinating seed, how the supply of food is gradually exhausted as the plant grows. And it lasts just long enough for the young plant to form roots and green leaves by which to obtain its nourishment direct from the soil and air. (See germination of Horse Gram p. 31, of Cucumber p. 73, and of Rice p. 169.)

APPENDIX

1. Nomenclature and Classification of Plants.

1. **Species.**—The fact that plants, reared from the seeds of a mother-plant, are like each other and like the parent, is familiar to everybody. Thus plants that spring from the seeds of a Banyan tree become Banyan trees again, and such as spring from Cucumber-seeds become Cucumbers. All such plants, therefore, which appear to have sprung from the same parent, and agree with each other in all essential parts, constitute a *species*.

There is, however, some variation in the development of the various parts of plants belonging to the same species, that is caused by differences of soil, climate, and other conditions. In identifying plants beginners must, therefore, distinguish between essential and accidental variations (*cf.* Tumbe, page 106). The following extract from Sir Jos. Hooker gives the chief causes of such variations:—

"A bright light and open situation, particularly at considerable elevations, without too much wet or drought, tends to increase the size and heighten the colour of flowers in proportion to the stature and foliage of the plant. Shade, on the contrary, especially with rich soil and sufficient moisture, tends to increase the foliage and draw up the stem, but to diminish the number, size, and colour of the flowers. A hot climate and dry situation tend to increase the hairs, prickles, and other productions of the epidermis, and to shorten and stiffen the branches. Moisture in a rich soil has the contrary effect. The neighbourhood of the sea, or a saline soil or atmosphere, imparts a thicker and more succulent consistence to the foliage and almost every part of the plant, and appears not unfrequently to enable plants, usually annual, to live through the winter.

"The luxuriance of plants growing in a rich soil, and the dwarf, stunted character of those crowded in poor soils, or in the cold, damp regions of high mountain-tops, is well known."

2. **Genera.**—We have seen that the Banyan tree *(Ficus bengalensis)* constitutes a species. If we muster the vegetable kingdom, we can easily find other trees which resemble the Banyan tree in most important points of structure, such as the Peepul tree or the Country Fig tree. Such plants form one *genus, viz.,* the genus "Ficus".

3. The **Scientific Nomenclature** of plants is based on this classification into species and genera. Thus the Banyan tree is known as *Ficus bengalensis* and the Peepul tree as *Ficus religiosa,* the common first name "Ficus" denoting the genus, and the second "bengalensis" or "religiosa", the species.

Plants have, of course, popular names also; but as these vary not only in various countries, but even in different parts of the same country, and as different plants are also called by the same name in different parts of a country, such popular names are useless for students of botany. Hence scientific names, derived from Latin and Greek, are applied to plants, by which they are known to all educated people of the world.

4. **Families, Classes, etc.**—Several genera which agree in certain marked characters, constitute a family, and several families still larger divisions. In this manner we have grouped:

the *genera* of Gossypium, Hibiscus, Bombax, etc., under the *family* of Malvaceæ;

the *families* of Malvaceæ, Cruciferæ, Leguminosæ and others under the *sub-class* of Polypetalæ (= plants with separated petals);

the *sub-classes* of polypetalous, gamopetalous (= of united petals) and monochlamydeous plants (= having a single instead of a double floral cover) under the *class* of Dicotyledons;

the *classes* of dicotyledonous and monocotyledonous plants under the *division* of Phanerogams (= "flowering plants");

and the two *divisions* of phanerogamous and cryptogamous ("flowerless") plants under the Vegetable *Kingdom.*

2. Distribution of Plants.

1. Any ramble round about our houses teaches us that plants are variously distributed. Some prefer the open field as their habitat, others the shady woods; some prefer the rich loam near rivers and tanks; others choose the dry and rocky soil of hills or deserts. It is **the difference of soil, light, and moisture,** which thus conditions the change of vegetation in various places.

2. An excursion into remoter parts of our country, or to the top of a high mountain, shows a still greater difference. This is due to **the change of the climate,** *viz.*, heat and moisture, in various parts of India.

So, when we ascend a mountain of the Himalayas, we have to travel through different belts of vegetation: about the base we find Palms and Bananas; a little higher, Bamboos, Figs, Ferns, etc.; higher up, Myrtles and Laurels; then Conifers and dwarf-trees; and in the highest regions the flowering plants cease to grow, and leave the place to Mosses and other Cryptogams.

The same succession of different classes of plants is noticed by travellers from the equatorial to the polar regions of the earth.

3. An important part in the distribution of plants is played **by man**. It is he who has brought new plants from foreign countries and from far remote continents to our land (Spanish Pepper, Coffee, Tobacco, Potato, and many others); who has cultivated large tracts of land with food-crops and other useful plants (Cereals, Pulses, Cotton, etc.), having suppressed the indigenous weeds that were growing there before; who cuts down forests to plant crops more necessary; who drains swamps and irrigates deserts to make them fertile.

3. Bibliography.

J. D. Hooker.—*Flora of British India.*. 1872—97, London, 7 vols.

W. Roxburgh.—*Flora Indica.* 1832, Serampore.

R. Wight.—*Icones Plantarum Indiæ Orientalis.* 1838—53, Madras, 6 vols.

D. Brandis.—*Indian Trees.* 1906, London.

A. F. W. Schimper.—*Pflanzengeographie.* 1898, Jena.

E. Strassburger, Noll, Schenck, and Karsten.—*Lehrbuch der Botanik.* 9th ed. 1908, Jena.

O. Schmeil.—*Lehrbuch der Botanik.* 10th ed. 1905, Stuttgart.

D. Oliver.—*First Book of Indian Botany.* 1889, London.

A. K. Nairne.—*The Flowering Plants of Western India.* 1894, London.

Basel Mission Press.—*Five Hundred Indian Plants*—in Kanarese. 1908, Mangalore.

Fyson.—*Indian Botany.* C. L. S., Madras.

INDEX TO VERNACULAR NAMES

INDEX